Ichthyology: An Introduction to Fish Science

Ichthyology: An Introduction to Fish Science

Edited by
Thomas Keenan

Larsen & Keller
www.larsen-keller.com

Ichthyology: An Introduction to Fish Science
Edited by Thomas Keenan
ISBN: 978-1-63549-762-5 (Hardback)

◫ Larsen & Keller

Published by Larsen and Keller Education,
5 Penn Plaza,
19th Floor,
New York, NY 10001, USA

Cataloging-in-Publication Data

Ichthyology : an introduction to fish science / edited by Thomas Keenan.
     p. cm.
Includes bibliographical references and index.
ISBN 978-1-63549-762-5
1. Ichthyology. 2. Fishes. 3. Fishery sciences. I. Keenan, Thomas.
QL615 .I24 2018
597--dc23

For more information regarding Larsen and Keller Education and its products, please visit the publisher's website www.larsen-keller.com

Table of Contents

Preface

Ichthyology is a branch of zoology which is concerned with the study of fishes, their biology, structure, organs and discovering their species. It includes species like chondrichthyes the cartilaginous fish, jawless fish i.e agnatha, and bony fish i.e osteichthyes. There are approximately 33,400 species of fishes being studied under ichthyology. The book aims to shed light on some of the unexplored aspects of this field. Such selected concepts that redefine ichthyology have been presented in this text. It unfolds the innovative aspects of this area which will be crucial for the holistic understanding of the subject matter. Those in search of information to further their knowledge will be greatly assisted by this textbook.

Given below is the chapter wise description of the book:

Chapter 1- Ichthyology is the study of fish. It includes various kinds of fishes like cartilaginous fish, jawless fish and bony fish. The evolution of fish is mainly focused upon in the subject of ichthyology. The chapter on ichthyology offers an insightful focus, keeping in mind the complex subject matter.

Chapter 2- Fishes can be classified into osteichthyes, chondrichthyes and agnatha. Chondrichthyes is a class of fish which can be further divided into elasmobranchii and holocephali. The topics discussed in the chapter are of great importance to broaden the existing knowledge on fishes.

Chapter 3- Fish physiology is the study of the parts and functions of fishes. Some of these parts are jaw, scale, gill and fin. Fish locomotion, electroreception and fish intelligence are some of the significant and important topics related to fish physiology. The following chapter unfolds its crucial aspects in a critical yet systematic manner.

Chapter 4- Fish farming is the raising of fishes for commercial purposes. The commonly farmed fishes are Atlantic salmon, sea bass, turbot and trout. This chapter discusses the methods of fish farming in a critical manner providing key analysis to the subject matter.

At the end, I would like to thank all those who dedicated their time and efforts for the successful completion of this book. I also wish to convey my gratitude towards my friends and family who supported me at every step.

Editor

An Introduction to Ichthyology

Ichthyology is the study of fish. It includes various kinds of fishes like cartilaginous fish, jawless fish and bony fish. The evolution of fish is mainly focused upon in the subject of ichthyology. The chapter on ichthyology offers an insightful focus, keeping in mind the complex subject matter.

Ichthyology

Coelacanth

Ichthyology also known as fish science, is the branch of zoology devoted to the study of fish. This includes bony fish (Osteichthyes), cartilaginous fish (Chondrichthyes), and jawless fish (Agnatha). While a large number of species have been discovered, approximately 250 new species are officially described by science each year. According to FishBase, 33,400 species of fish had been described by October 2016.

History

The study of fish receives its origins from human's desire to feed, clothe, and equip themselves with useful implements. According to Michael Barton, a prominent ichthyologist and professor at Centre College, "the earliest ichthyologists were hunters and gatherers who had learned how to obtain the most useful fish, where to obtain them in abundance, and at what times they might be the most available". Early cultures manifested these insights in abstract and identifiable artistic expressions.

The study of fish dates from the Upper Paleolithic Revolution (with the advent of "high culture"). The science of ichthyology was developed in several interconnecting epochs, each with various significant advancements.

1500 BC–40 AD

Informal, scientific descriptions of fish are represented within the Judeo-Christian tradition. The Old Testament laws of kashrut forbade the consumption of fish without scales or appendages. Theologians and ichthyologists believe that the apostle Peter and his contemporaries harvested the fish that are today sold in modern industry along the Sea of Galilee, presently known as Lake

Kinneret. These fish include cyprinids of the genera *Barbus* and *Mirogrex*, cichlids of the genus *Sarotherodon*, and *Mugil cephalus* of the family Mugilidae.

Fish represent approximately 8% of all figurative depictions on Mimbres pottery.

335 BC–80 AD

Aristotle incorporated ichthyology into formal scientific study. Between 335 BC–322 BC, he provided the earliest taxonomic classification of fish, accurately describing 117 species of Mediterranean fish. Furthermore, Aristotle documented anatomical and behavioral differences between fish and marine mammals. After his death, some of his pupils continued his ichthyological research. Theophrastus, for example, composed a treatise on amphibious fish. The Romans, although less devoted to science, wrote extensively about fish. Pliny the Elder, a notable Roman naturalist, compiled the ichthyological works of indigenous Greeks, including verifiable and ambiguous peculiarities such as the sawfish and mermaid respectively. Pliny's documentation was the last significant contribution to ichthyology until the European Renaissance.

European Renaissance

The writings of three sixteenth century scholars, Hippolito Salviani, Pierre Belon, and Guillaume Rondelet, signify the conception of modern ichthyology. The investigations of these individuals were based upon actual research in comparison to ancient recitations. This property popularized and emphasized these discoveries. Despite their prominence, Rondelet's *De Piscibus Marinis* is regarded as the most influential, identifying 244 species of fish.

16th–17th Century

The incremental alterations in navigation and shipbuilding throughout the Renaissance marked the commencement of a new epoch in ichthyology. The Renaissance culminated with the era of exploration and colonization, and upon the cosmopolitan interest in navigation came the specialization in naturalism. Georg Marcgrave of Saxony composed the *Naturalis Brasilae* in 1648. This document contained a description of 100 species of fish indigenous to the Brazilian coastline. In 1686, John Ray and Francis Willughby collaboratively published *Historia Piscium*, a scientific manuscript containing 420 species of fish, 178 of these newly discovered. The fish contained within this informative literature were arranged in a provisional system of classification.

PETRI ARTEDI·

SVECI, MEDICI

ICHTHYOLOGIA

SIVE

OPERA OMNIA

PISCIBUS

SCILICET:

BIBLIOTHECA ICHTHYOLOGICA,
PHILOSOPHIA ICHTHYOLOGICA,
GENERA PISCIUM.
SYNONYMIA SPECIERUM.
DESCRIPTIONES SPECIERUM.

OMNIA IN HOC GENERE PERFECTIORA,
QUAM ANTEA ULLA.

POSTHUMA

Vindicavit, Recognovit, Coaptavit & Edidit

CAROLUS LINNÆUS,

Med. Doct. & Ac. Imper. N. C.

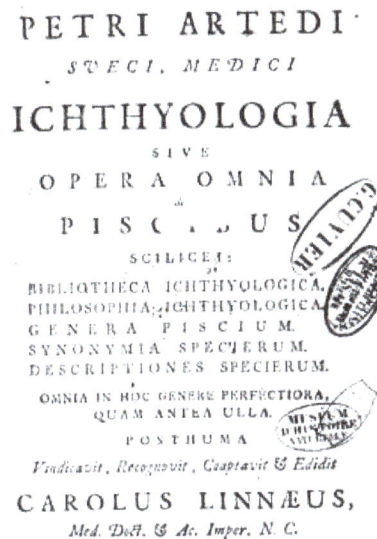

Frontipiece from *Ichthyologia, sive Opera Omnia de Piscibus* by Peter Artedi

The classification used within the *Historia Piscium* was further developed by Carl Linnaeus, the "father of modern taxonomy". His taxonomic approach became the systematic approach to the study of organisms, including fish. Linnaeus was a professor at the University of Uppsala and an eminent botanist; however, one of his colleagues, Peter Artedi, earned the title "father of ichthyology" through his indispensable advancements. Artedi contributed to Linnaeus's refinement of the principles of taxonomy. Furthermore, he recognized five additional orders of fish: Malacopterygii, Acanthopterygii, Branchiostegi, Chondropterygii, and Plagiuri. Artedi developed standard methods for making counts and measurements of anatomical features that are modernly exploited. Another associate of Linnaeus, Albertus Seba, was a prosperous pharmacist from Amsterdam. Seba assembled a cabinet, or collection, of fish. He invited Artedi to utilize this assortment of fish; unfortunately, in 1735, Artedi fell into an Amsterdam canal and drowned at the age of 30.

Linnaeus posthumously published Artedi's manuscripts as *Ichthyologia, sive Opera Omnia de Piscibus* (1738). His refinement of taxonomy culminated in the development of the binomial nomenclature which is in use by contemporary ichthyologists. Furthermore, he revised the orders introduced by Artedi, placing significance on pelvic fins. Fish lacking this appendage were placed within the order Apodes; fish containing abdominal, thoracic, or jugular pelvic fins were termed Abdominales, Thoracici, and Jugulares respectively. However, these alterations were not grounded within evolutionary theory. Therefore, it would take over a century until Charles Darwin would provide the intellectual foundation from which we would be permitted to perceive that the degree of similarity in taxonomic features was a consequence of phylogenetic relationship.

Modern Era

Close to the dawn of the nineteenth century, Marcus Elieser Bloch of Berlin and Georges Cuvier of Paris made attempts to consolidate the knowledge of ichthyology. Cuvier summarized all of the available information in his monumental *Histoire Naturelle des Poissons*. This manuscript was published between 1828 and 1849 in a 22 volume series. This documental describes 4,514 species of fish, 2,311 of these new to science. It remains one of the most ambitious treatises of the modern world. Scientific exploration of the Americas advanced our knowledge of the remarkable diversity of fish. Charles Alexandre Lesueur was a student of Cuvier. He made a cabinet of fish dwelling within the Great Lakes and Saint Lawrence River regions.

Adventurous individuals such as John James Audubon and Constantine Samuel Rafinesque figure in the faunal documentation of North America. These persons often traveled with one another. Rafinesque wrote *Ichthyologia Ohiensis* in 1820. In addition, Louis Agassiz of Switzerland established his reputation through the study of freshwater fish and the first comprehensive treatment of paleoichthyology, Poissons Fossiles. In the 1840s, Agassiz moved to the United States, where he taught at Harvard University until his death in 1873.

Albert Günther published his *Catalogue of the Fish of the British Museum* between 1859 and 1870, describing over 6,800 species and mentioning another 1,700. Generally considered one of the most influential ichthyologists, David Starr Jordan wrote 650 articles and books on the subject as well as serving as president of Indiana University and Stanford University.

Evolution of Fish

The evolution of fish began about 530 million years ago during the Cambrian explosion. It was during this time that the early chordates developed the skull and the vertebral column, leading to the first craniates and vertebrates. The first fish lineages belong to the Agnatha, or jawless fish. Early examples include *Haikouichthys*. During the late Cambrian, eel-like jawless fish called the conodonts, and small mostly armoured fish known as ostracoderms, first appeared. Most jawless fish are now extinct; but the extant lampreys may approximate ancient pre-jawed fish. Lampreys belong to the Cyclostomata, which includes the extant hagfish, and this group may have split early on from other agnathans.

The first jawed vertebrates probably developed during the late Ordovician period. They are first represented in the fossil record from the Silurian by two groups of fish: the armoured fish known as placoderms, which evolved from the ostracoderms; and the Acanthodii (or spiny sharks). The jawed fish that are still extant in modern days also appeared in late Silurian: the Chondrichthyes (or cartilaginous fish) and the Osteichthyes (or bony fish). The bony fish evolved into two separate groups: the Actinopterygii (or ray-finned fish) and Sarcopterygii (which includes the lobe-finned fish).

During the Devonian period a great increase in fish variety occurred, especially that of the ostracoderms and placoderms, as well as lobe-finned fish and early sharks. This has led to the Devonian being known as the *age of fishes*. It was from the lobe-finned fish that the tetrapods evolved, the non-fish four-limbed vertebrates, represented today by amphibians, mammals, reptiles and birds. Transitional tetrapods first appeared during the early Devonian, and by the late Devonian the first tetrapods appeared. The diversity of jawed vertebrates may indicate the evolutionary advantage of a jawed mouth; but it is unclear if the advantage of a hinged jaw is greater biting force, improved respiration, or a combination of factors. Fish do not represent a monophyletic group, but a paraphyletic one, as they exclude the tetrapods.

Fish, like many other organisms, have been greatly affected by extinction events throughout natural history. The Ordovician–Silurian extinction events led to the loss of many species. The late Devonian extinction led to the extinction of the ostracoderms and placoderms by the end of the Devonian, as well as other fish. The spiny sharks became extinct at the Permian–Triassic extinction event; the conodonts became extinct at the Triassic–Jurassic extinction event. The Cretaceous–Paleogene extinction event, and the present day Holocene extinction, have also affected fish variety and fish stocks.

Overview

Vertebrate Classes

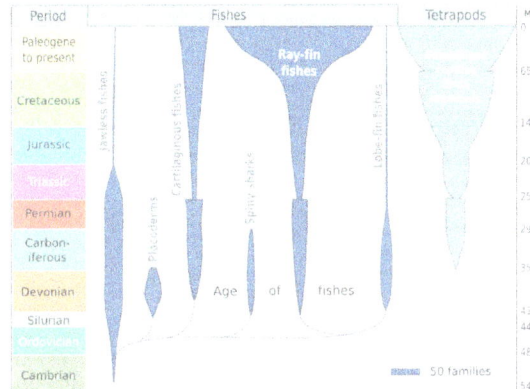

Spindle diagram for the evolution of fish and other vertebrate classes.
The diagram is based on Michael Benton, 2005.

Conventional classification has living vertebrates as a subphylum grouped into eight classes based on traditional interpretations of gross anatomical and physiological traits. In turn, these classes are grouped into the vertebrates that have four limbs (the tetrapods) and those that do not: fishes. The extant vertebrate classes are:

Fish:

- jawless fishes (Agnatha)
- cartilaginous fishes (Chondrichthyes)
- ray-finned fishes (Actinopterygii)
- lobe-finned fishes (Sarcopterygii)

Tetrapods:

- amphibians (Amphibia)
- reptiles (Reptilia)
- birds (Aves)
- mammals (Mammalia)

In addition to these are two classes of extinct jawed fishes, the armoured placoderms and the spiny sharks.

Fish may have evolved from an animal similar to a coral-like sea squirt (a tunicate), whose larvae resemble early fish in important ways. The first ancestors of fish may have kept the larval form into adulthood (as some sea squirts do today), although perhaps the reverse is the case. Vertebrates, among them the first fishes, originated about 530 million years ago during the Cambrian explosion, which saw the rise in organism diversity.

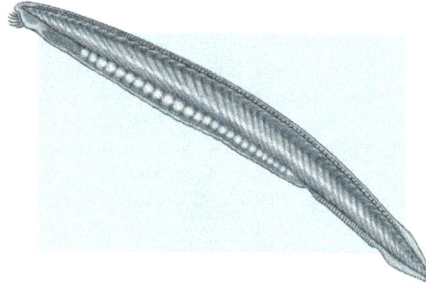

The lancelet, a small, translucent, fish-like animal, is the closest living invertebrate relative of the olfactoreans (vertebrates and tunicates).

The early vertebrate *Haikouichthys*, from about 518 million years ago in China, may be the "ancestor to all vertebrates" and is one of the earliest known fish.

The first ancestors of fish, or animals that were probably closely related to fish, were *Pikaia*, *Haikouichthys* and *Myllokunmingia*. These three genera all appeared around 530 Ma. *Pikaia* had a primitive notochord, a structure that could have developed into a vertebral column later. Unlike the other fauna that dominated the Cambrian, these groups had the basic vertebrate body plan: a notochord, rudimentary vertebrae, and a well-defined head and tail. All of these early vertebrates lacked jaws in the common sense and relied on filter feeding close to the seabed.

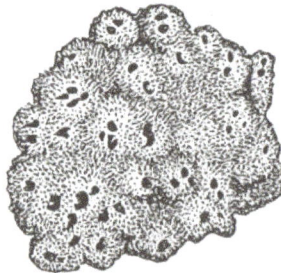

Somewhat dated view of continuous evolutionary gradation

These were followed by indisputable fossil vertebrates in the form of heavily armoured fishes discovered in rocks from the Ordovician Period 500–430 Ma.

The first jawed vertebrates appeared in the late Ordovician and became common in the Devonian, often known as the "Age of Fishes". The two groups of bony fishes, the actinopterygii and sarcopterygii, evolved and became common. The Devonian also saw the demise of virtually all jawless fishes, save for lampreys and hagfish, as well as the Placodermi, a group of armoured fish that dominated much of the late Silurian. The Devonian also saw the rise of the first labyrinthodonts, which was a transitional between fishes and amphibians.

The colonisation of new niches resulted in diversification of body plans and sometimes an increase in size. The Devonian Period (395 to 345 Ma) brought in such giants as the placoderm *Dunkleosteus*, which could grow up to seven meters long, and early air-breathing fish that could remain on land for extended periods. Among this latter group were ancestral amphibians.

The reptiles appeared from labyrinthodonts in the subsequent Carboniferous period. The anapsid and synapsid reptiles were common during the late Paleozoic, while the diapsids became dominant during the Mesozoic. In the sea, the bony fishes became dominant.

The later radiations, such as those of fish in the Silurian and Devonian periods, involved fewer taxa, mainly with very similar body plans. The first animals to venture onto dry land were arthropods. Some fish had lungs and strong, bony fins and could crawl onto the land also.

Jawless Fish

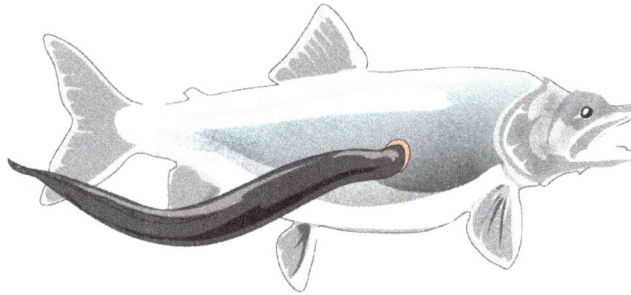

A modern jawless fish, the lamprey, attached to a modern jawed fish

Lamprey mouth

Jawless fishes belong to the superclass Agnatha in the phylum Chordata, subphylum Vertebrata. Agnatha comes from the Greek, and means "no jaws". It excludes all vertebrates with jaws, known as gnathostomes. Although a minor element of modern marine fauna, jawless fish were prominent among the early fish in the early Paleozoic. Two types of Early Cambrian animal apparently having fins, vertebrate musculature, and gills are known from the early Cambrian Maotianshan shales of China: *Haikouichthys* and *Myllokunmingia*. They have been tentatively assigned to Agnatha by Janvier. A third possible agnathid from the same region is *Haikouella*. A possible agnathid that has not been formally described was reported by Simonetti from the Middle Cambrian Burgess Shale of British Columbia.

Many Ordovician, Silurian, and Devonian agnathians were armoured with heavy bony-spiky plates. The first armoured agnathans—the Ostracoderms, precursors to the bony fish and hence to the tetrapods (including humans)—are known from the middle Ordovician, and by the Late Silurian the agnathans had reached the high point of their evolution. Most of the ostracoderms, such as thelodonts, osteostracans, and galeaspids, were more closely related to the gnathostomes than to the surviving agnathans, known as cyclostomes. Cyclostomes apparently split from other agnathans before the evolution of dentine and bone, which are present in many fossil agnathans, including conodonts. Agnathans declined in the Devonian and never recovered.

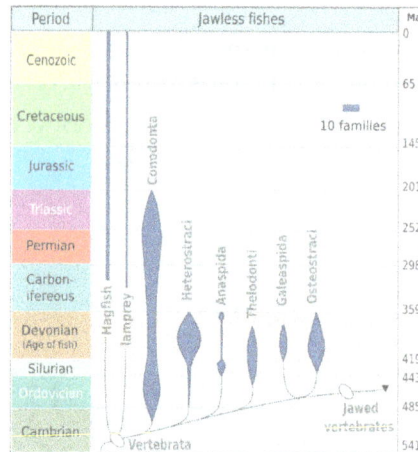

Evolution of jawless fishes. The diagram is based on Michael Benton, 2005.

The agnathans as a whole are paraphyletic, because most extinct agnathans belong to the stem group of gnathostomes. Recent molecular data, both from rRNA and from mtDNA strongly supports the theory that living agnathans, known as cyclostomes, are monophyletic. In phylogenetic taxonomy, the relationships between animals are not typically divided into ranks, but illustrated as a nested "family tree" known as a cladogram. Phylogenetic groups are given definitions based on their relationship to one another, rather than purely on physical traits such as the presence of a backbone. This nesting pattern is often combined with traditional taxonomy, in a practice known as evolutionary taxonomy.

The cladogram below for jawless fish is based on studies compiled by Philippe Janvier and others for the *Tree of Life Web Project*. († = group is extinct)

†Conodonts

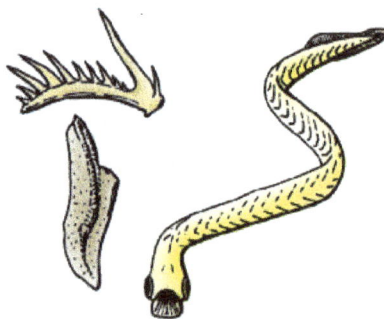

†Conodonts (extinct) resembled primitive jawless eels

Conodonts resembled primitive jawless eels. They appeared 495 Ma and were wiped out 200 Ma. Initially they were known only from tooth-like microfossils called *conodont elements*. These "teeth" have been variously interpreted as filter-feeding apparatuses or as a "grasping and crushing array". Conodonts ranged in length from a centimeter to the 40 cm *Promissum*. Their large eyes had a lateral position, which makes a predatory role unlikely. The preserved musculature hints that some conodonts (*Promissum* at least) were efficient cruisers but incapable of bursts of speed. In 2012 researchers classify the conodonts in the phylum Chordata on the basis of their fins with fin rays, chevron-shaped muscles and notochord. Some researchers see them as vertebrates similar in appearance to modern hagfish and lampreys, though phylogenetic analysis suggests that they are more derived than either of these groups.

†Ostracoderms

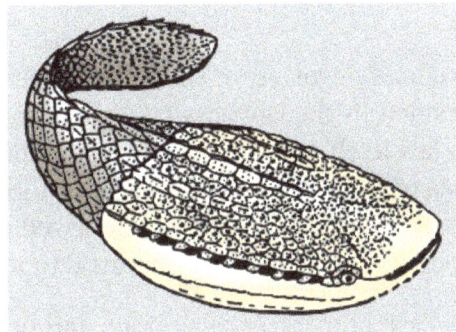

†Ostracoderms (extinct) were armoured jawless fishes

Ostracoderms *(shell-skinned)* are armoured jawless fishes of the Paleozoic. The term does not often appear in classifications today because it is paraphyletic or polyphyletic, and has no phylogenetic meaning. However, the term is still used informally to group together the armoured jawless fishes.

The ostracoderm armour consisted of 3–5 mm polygonal plates that shielded the head and gills, and then overlapped further down the body like scales. The eyes were particularly shielded. Earlier chordates used their gills for both respiration and feeding, whereas ostracoderms used their gills for respiration only. They had up to eight separate pharyngeal gill pouches along the side of the head, which were permanently open with no protective operculum. Unlike invertebrates that use ciliated motion to move food, ostracoderms used their muscular pharynx to create a suction that pulled small and slow moving prey into their mouths.

The first fossil fishes that were discovered were ostracoderms. The Swiss anatomist Louis Agassiz received some fossils of bony armored fish from Scotland in the 1830s. He had a hard time classifying them as they did not resemble any living creature. He compared them at first with extant armored fish such as catfish and sturgeons but later realizing that they had no movable jaws, classified them in 1844 into a new group "ostracoderms".

Ostracoderms existed in two major groups, the more primitive heterostracans and the cephalaspids. Later, about 420 million years ago, the jawed fish evolved from one of the ostracoderms. After the appearance of jawed fish, most ostracoderm species underwent a decline, and the last ostracoderms became extinct at the end of the Devonian period.

Jawed Fish

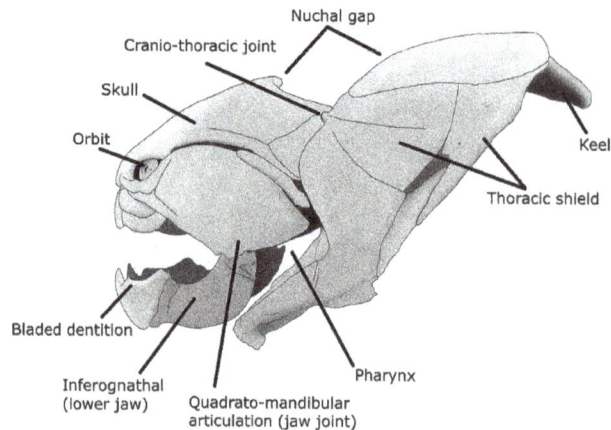

The vertebrate jaw probably originally evolved in the Silurian period and appeared in the Plac-oderm fish, which further diversified in the Devonian. The two most anterior pharyngeal arches are thought to have become the jaw itself and the hyoid arch, respectively. The hyoid system sus-pends the jaw from the braincase of the skull, permitting great mobility of the jaws. Already long assumed to be a paraphyletic assemblage leading to more derived gnathostomes, the discovery of *Entelognathus* suggests that placoderms are directly ancestral to modern bony fish.

As in most vertebrates, fish jaws are bony or cartilaginous and oppose vertically, comprising an *upper jaw* and a *lower jaw*. The jaw is derived from the most anterior two pharyngeal arches supporting the gills, and usually bears numerous teeth. The skull of the last common ancestor of today's jawed vertebrates is assumed to have resembled sharks.

Jawed fish and Evolution of jaws

It is thought that the original selective advantage garnered by the jaw was not related to feeding, but to increased respiration efficiency. The jaws were used in the buccal pump (observable in mod-ern fish and amphibians) that pumps water across the gills of fish or air into the lungs in the case of amphibians. Over evolutionary time the more familiar use of jaws (to humans), in feeding, was selected for and became a very important function in vertebrates. Many teleost fish have substan-tially modified their jaws for suction feeding and jaw protrusion, resulting in highly complex jaws with dozens of bones involved.

Jawed vertebrates and jawed fish evolved from jawless fish, and the cladogram below for jawed vertebrates is a continuation of the cladogram in the section above. († = group is extinct)

†Placoderms

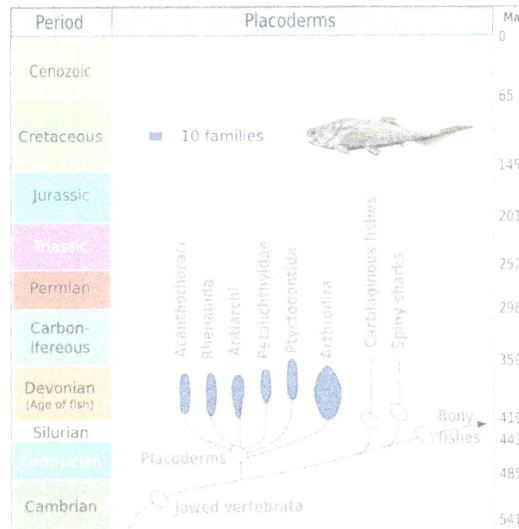

Period	Placoderms	Ma
Cenozoic		0
		65
Cretaceous	10 families	
		145
Jurassic		
		201
Triassic		252
Permian		298
Carbon-iferous		
		359
Devonian (Age of fish)		
Silurian		419
		443
Ordovician		485
Cambrian	Jawed vertebrata	541

Evolution of the (now extinct) placoderms. The diagram is
based on Michael Benton, 2005.

Placoderms, class Placodermi *(plate skinned)*, are extinct armoured prehistoric fish, which appeared about 430 Ma in the Early to Middle Silurian. They were mostly wiped out during the Late Devonian Extinction event, 378 Ma, though some survived and made a slight recovery in diversity during the Famennian epoch before dying out entirely at the close of the Devonian, 360 mya; they are ultimately ancestral to modern vertebrates. Their head and thorax were covered with massive and often ornamented armoured plates. The rest of the body was scaled or naked, depending on the species. The armour shield was articulated, with the head armour hinged to the thoratic armour. This allowed placoderms to lift their heads, unlike ostracoderms. Placoderms were the first jawed fish; their jaws likely evolved from the first of their gill arches. The chart on the right shows the rise and demise of the separate placoderm lineages: Acanthothoraci, Rhenanida, Antiarchi, Petalichthyidae, Ptyctodontida and Arthrodira.

†Placoderms (extinct) were armoured jawed fishes
(compare with the ostracoderms above)

†Spiny Sharks

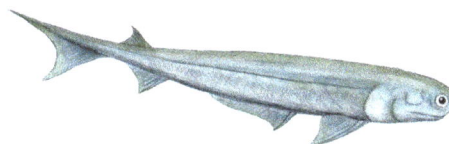

†Spiny sharks (extinct) were the earliest known jawed fishes. They
resembled sharks and were indeed ancestral to them.

Spiny sharks, class Acanthodii, are extinct fishes that share features with both bony and cartilaginous fishes, though ultimately more closely related to and ancestral to the latter. Despite being called "spiny sharks", acanthodians predate sharks, though they gave rise to them. They evolved in the sea at the beginning of the Silurian Period, some 50 million years before the first sharks appeared. Eventually competition from bony fishes proved too much, and the spiny sharks died out in Permian times about 250 Ma. In form they resembled sharks, but their epidermis was covered with tiny rhomboid platelets like the scales of holosteans (gars, bowfins).

Cartilaginous Fishes

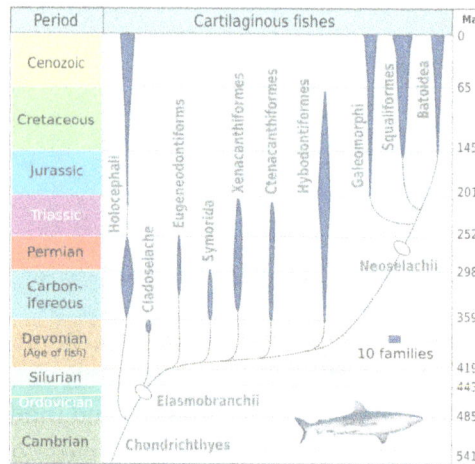

Radiation of cartilaginous fishes, derived from work by Michael Benton, 2005.

Cartilaginous fishes, class Chondrichthyes, consisting of sharks, rays and chimaeras, appeared by about 395 million years ago, in the middle Devonian, evolving from acanthodians. The class contains the sub classes Holocephali (chimaera) and Elasmobranchii (sharks and rays). The radiation of elasmobranches in the chart on the right is divided into the taxa: Cladoselache, Eugeneodontiformes, Symmoriida, Xenacanthiformes, Ctenacanthiformes, Hybodontiformes, Galeomorphi, Squaliformes and Batoidea.

Bony Fishes

Bony fishes, class Osteichthyes, are characterised by bony skeleton rather than cartilage. They appeared in the late Silurian, about 419 million years ago. The recent discovery of *Entelognathus* strongly suggests that bony fishes (and possibly cartilaginous fishes, via acanthodians) evolved from early placoderms. A subclass of the Osteichthyes, the ray-finned fishes (Actinopterygii), have become the dominant group of fishes in the post-Paleozoic and modern world, with some 30,000 living species.

The bony (and cartilaginous) fish groups that emerged after the Devonian, were characterised by steady improvements in foraging and locomotion.

Lobe-finned Fishes

Lobe-finned fishes, fish belonging to the class Sarcopterygii, are mostly extinct bony fishes, basally characterised by robust and stubby lobe fins containing a robust internal skeleton, cosmoid scales and internal nostrils. Their fins are fleshy, lobed, paired fins, joined to the body by a single bone. The fins of lobe-finned fish differ from those of all other fish in that each is borne on a fleshy, lobelike, scaly stalk extending from the body. The pectoral and pelvic fins are articulated in ways resembling the tetrapod limbs they were the precursors to. The fins evolved into the legs of the first tetrapod land vertebrates, amphibians. They also possess two dorsal fins with separate bases, as opposed to the single dorsal fin of ray-finned fish. The braincase of lobe-finned fishes primitively has a hinge line, but this is lost in tetrapods and lungfish. Many early lobe-finned fishes have a symmetrical tail. All lobe-finned fishes possess teeth covered with true enamel.

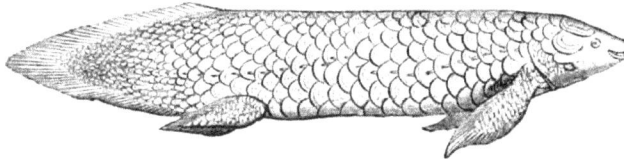

The Queensland lungfish is a lobe-finned fish that is a *living fossil*. Lungfish evolved the first proto-lungs and proto-limbs. They developed the ability to live outside a water environment in the middle Devonian (397-385 Ma), and have remained virtually the same for over 100 million years. Phylogenomic analysis has shown that "the closest living fish to the tetrapod ancestor is the lungfish, not the coelacanth".

Lobe-finned fishes, such as coelacanths and lungfish, were the most diverse group of bony fishes in the Devonian. Taxonomists who subscribe to the cladistic approach include the grouping Tetrapoda within the Sarcopterygii, and the tetrapods in turn include all species of four-limbed vertebrates. The fin-limbs of lobe-finned fishes such as the coelacanths show a strong similarity to the expected ancestral form of tetrapod limbs. The lobe-finned fish apparently followed two different lines of development and are accordingly separated into two subclasses, the Rhipidistia (including the lungfish, and the Tetrapodomorpha, which include the Tetrapoda) and the Actinistia (coelacanths). The first lobe-finned fishes, found in the uppermost Silurian (ca 418 Ma), closely resembled spiny sharks, which became extinct at the end of the Paleozoic. In the early–middle Devonian (416 - 385 Ma), while the predatory placoderms dominated the seas, some lobe-finned fishes came into freshwater habitats.

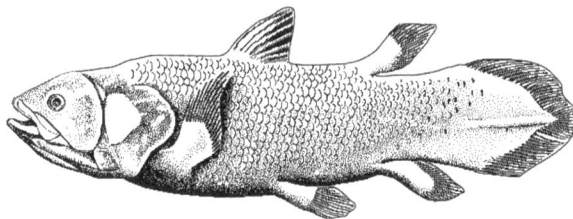

The coelacanth is another lobe-finned fish that is a living fossil. It is thought to have evolved into roughly its current form about 408 million years ago, during the early Devonian, and has not essentially evolved further from its ancient form.

In the Early Devonian (416-397 Ma), the lobe-finned fishes split into two main lineages — the coelacanths and the rhipidistians. The former never left the oceans and their heyday was the Late Devonian and Carboniferous, from 385 to 299 Ma, as they were more common during those periods than in any other period in the Phanerozoic; coelacanths still live today in the oceans (genus *Latimeria*). The Rhipidistians, whose ancestors probably lived in estuaries, migrated into freshwater habitats. They in turn split into two major groups: the lungfish and the tetrapodomorphs. The lungfish's greatest diversity was in the Triassic period; today there are fewer than a dozen genera left. The lungfish evolved the first proto-lungs and proto-limbs; developing the ability to live outside a water environment in the middle Devonian (397-385 Ma). The first tetrapodomorphs, which included the gigantic rhizodonts, had the same general anatomy as the lungfish, who were their closest kin, but they appear not to have left their water habitat until the late Devonian epoch (385 - 359 Ma), with the appearance of tetrapods (four-legged vertebrates). Tetrapods are the only tetrapodomorphs that survived after the Devonian. Lobe-finned fishes continued until towards the end of Paleozoic era, suffering heavy losses during the Permian-Triassic extinction event (251 Ma).

Ray-finned Fishes

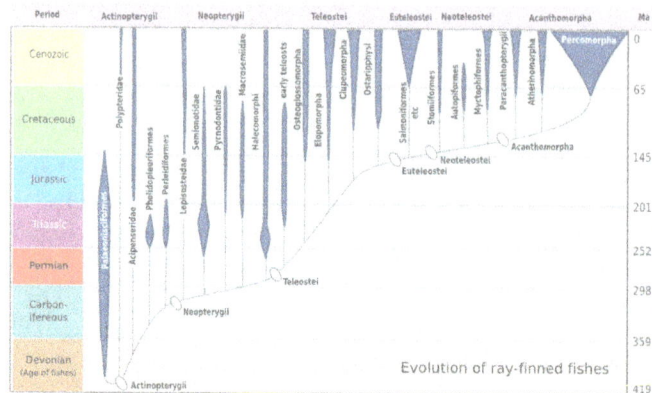

Evolution of ray-finned fishes

Ray-finned fishes, class Actinopterygii, differ from lobe-finned fishes in that their fins consist of webs of skin supported by spines ("rays") made of bone or horn. There are other differences in respiratory and circulatory structures. Ray-finned fishes normally have skeletons made from true bone, though this is not true of sturgeons and paddlefishes.

Ray-finned fishes are the dominant vertebrate group, containing half of all known vertebrate species. They inhabit abyssal depths in the sea, coastal inlets and freshwater rivers and lakes, and are a major source of food for humans.

Timeline

Pre Devonian: Origin of Fish

Cambrian (541–485 Ma): The beginning of the Cambrian was marked by the Cambrian explosion, the sudden appearance of nearly all of the invertebrate animal phyla (molluscs, jellyfish, worms and arthropods, such as crustaceans) in great abundance. The first vertebrates appeared in the form of primitive fish, which were subsequently greatly diversified in the Silurian and Devonian.

Pikaia

Pikaia, along with Myllokunmingia and Haikouichthys ercaicunensis immediately below, are all candidates in the fossil record for the titles of "first vertebrate" and "first fish". Pikaia is a genus that appeared about 530 Ma during the Cambrian explosion of multicellular life. Pikaia gracilens (pictured) is a transitional fossil between invertebrates and vertebrates, and may be the earliest known chordate. In this sense it may have been the original ancestor of fishes. It was a primitive creature with no evidence of eyes, without a well defined head, and less than 2 inches (5 centimetres) long. Pikaia was a sideways-flattened, leaf-shaped animal that swam by throwing its body into a series of S-shaped, zig-zag curves, similar to movement of snakes. Fish inherited the same swimming movement, but they generally have stiffer backbones. It had a pair of large head tentacles and a series of short appendages, which may be linked to gill slits, on either side of its head. Pikaia shows the essential prerequisites for vertebrates. The flattened body is divided into pairs of segmented muscle blocks, seen as faint vertical lines. The muscles lie on either side of a flexible structure resembling a rod that runs from the tip of the head to the tip of the tail.

Pikaia

Haikouichthys

Haikouichthys (fish from Haikou) is another genus that also appears in the fossil record about 530 Ma, and also marks the transition from invertebrate to vertebrates. Haikouichthys are craniates (animals with backbones and distinct heads). Unlike Pikaia, they had eyes. They also had a defined skull and other characteristics that have led paleontologists to label it a true craniate, and even to be popularly characterized as one of the earliest fishes. Cladistic analysis indicates that the animal is probably a basal chordate or a basal craniate; but it does not possess sufficient features to be included uncontroversially even in either stem group.

Haikouichthys

Myllokunmingia

Myllokunmingia is a genus that appeared about 530 Ma. It is a chordate, and it has been argued that it is a vertebrate, It is 28 mm long and 6 mm high, and is among the oldest possible craniates.

Conodont

Conodonts (cone-teeth) resembled primitive eels. They appeared 495 Ma and were wiped out 200

Ma. Initially they were known only from tooth-like microfossils called conodont elements. These "teeth" have been variously interpreted as filter-feeding apparatuses or as a "grasping and crushing array". Conodonts ranged in length from a centimeter to the 40 cm Promissum. Their large eyes had a lateral position of which makes a predatory role unlikely. The preserved musculature hints that some conodonts (Promissum at least) were efficient cruisers but incapable of bursts of speed. In 2012 researchers classify the conodonts in the phylum Chordata on the basis of their fins with fin rays, chevron-shaped muscles and notochord. Some researchers see them as vertebrates similar in appearance to modern hagfish and lampreys, though phylogenetic analysis suggests that they are more derived than either of these groups.

Ostracoderms

Ostracoderms (shell-skinned) are any of several groups of extinct, primitive, jawless fishes that were covered in an armour of bony plates. They appeared in the Cambrian, about 510 million years ago, and became extinct towards the end of the Devonian, about 377 million years ago. Initially Ostracoderms had poorly formed fins, and paired fins, or limbs, first evolved within this group. They were covered with a bony armour or scales and were often less than 30 cm (0.98 ft) long.

Ordovician (485–443 Ma): Fish, the world's first true vertebrates, continued to evolve, and those with jaws (Gnathostomata) may have first appeared late in this period. Life had yet to diversify on land.

Arandaspis

Arandaspis are jawless fish that lived in the early Ordovician period, about 480–470 Ma. It was about 15 cm (6 in) long, with a streamlined body covered in rows of knobbly armoured scutes. The front of the body and the head were protected by hard plates with openings for the eyes, nostrils and gills. Although it was jawless, Arandaspis might have had some moveable plates in its mouth, serving as lips, sucking in food particles. The low position of its mouth suggests it foraged the ocean floor. It lacked fins and its only method of propulsion was its horizontally flattened tail. As a result, it probably swam in a fashion similar to a modern tadpole.

Arandaspis

Ordovician (485–443 Ma): Fish, the world's first true vertebrates, continued to evolve, and those with jaws (Gnathostomata) may have first appeared late in this period. Life had yet to diversify on land.

Astraspis

Astraspis (star shield) is an extinct genus of primitive jawless fish related to other Ordovician fishes, such as Sacabambaspis and Arandaspis. Fossils show clear evidence of a sensory structure (lateral line system). The arrangement of these organs in regular lines allows the fish to detect the direction and distance from which a disturbance in the water is coming. Arandaspis are thought to have had a mobile tail covered with small protective plates and a head region covered with larger plates. A specimen described by Sansom et al. had relatively large, lateral eyes and a series of eight gill openings on each side.

Pteraspidomorphi

Pteraspidomorphi is an extinct class of early jawless fish. The fossils show extensive shielding of the head. Many had hypocercal tails to generate lift to increase ease of movement through the water for their armoured bodies, which were covered in dermal bone. They also had sucking mouth parts and some species may have lived in fresh water.

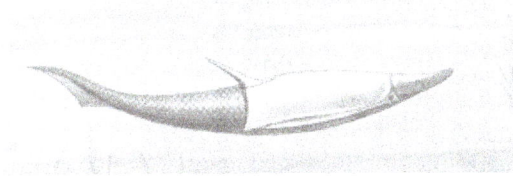

Pteraspidomorphi

Thelodonts

Thelodonts (nipple teeth) are a class of small, extinct jawless fishes with distinctive scales instead of large plates of armour. There is debate over whether these represent a monophyletic grouping, or disparate stem groups to the major lines of jawless and jawed fish. Thelodonts are united by their characteristic "thelodont scales". This defining character is not necessarily a result of shared ancestry, as it may have been evolved independently by different groups. Thus the thelodonts are generally thought to represent a polyphyletic group. If they are monophyletic, there is no firm evidence on what their ancestral state was. These scales were easily dispersed after death; their small size and resilience makes them the most common vertebrate fossil of their time. The fish lived in both freshwater and marine environments, first appearing during the Ordovician, and perishing during the Frasnian–Famennian extinction event of the Late Devonian. They were predominantly deposit-feeding bottom dwellers, although some species may have been pelagic.

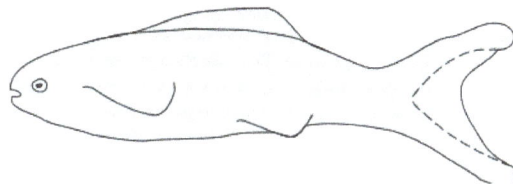

Thelodonts

The Ordovician ended with the Ordovician–Silurian extinction event (450–440 Ma). Two events occurred that killed off 27% of all families, 57% of all genera and 60% to 70% of all species. Together they are ranked by many scientists as the second largest of the five major extinctions in Earth's history in terms of percentage of genera that became extinct.

Silurian

Silurian (443–419 Ma): Many evolutionary milestones occurred during this period, including the appearance of armoured jawless fish, jawed fish, spiny sharks and ray-finned fish.

While it is traditional to refer to the Devonian as the age of fishes, recent findings have shown the Silurian was also a period of considerable diversification. Jawed fish developed movable jaws, adapted from the supports of the front two or three gill arches.

Anaspida

Anaspida (without shield) is an extinct class of primitive jawless vertebrates that lived during the Silurian and Devonian periods. They are classically regarded as the ancestors of lampreys. Anaspids were small, primarily marine agnathans that lacked heavy bony shield and paired fins, but have highly exaggerated hypocercal tails. They first appeared in the early Silurian, and flourished until the Late Devonian extinction, where most species, save for lampreys, became extinct. Unusually for an agnathan, anaspids did not possess a bony shield or armour. The head is instead covered in an array of smaller, weakly mineralised scales.

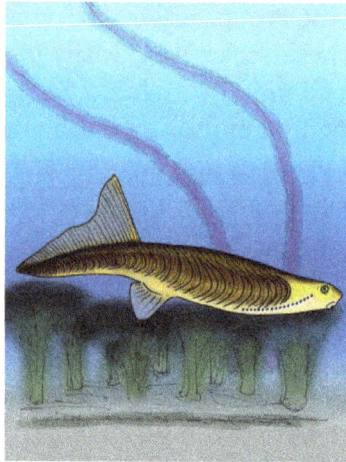

Anaspida

Osteostraci

Osteostraci ("bony shields") was a class of bony-armored jawless fish that lived from the Middle Silurian to Late Devonian. Anatomically speaking, the osteostracans, especially the Devonian species, were among the most advanced of all known agnathans. This is due to the development of paired fins, and their complicated cranial anatomy. The osteostracans were more similar to lampreys than to jawed vertebrates in possessing two pairs of semicircular canals in the inner ear, as opposed to the three pairs found in the inner ears of jawed vertebrates. Most osteostracans had a massive cephalothorac shield, but all Middle and Late Devonian species appear to have had a reduced, thinner, and often micromeric dermal skeleton. They were probably relatively good swimmers, possessing dorsal fins, paired pectoral fins, and a strong tail.

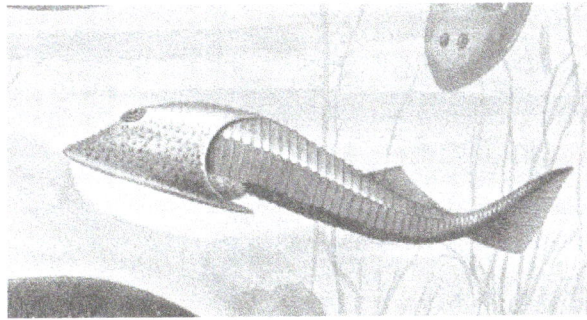
Osteostraci

Spiny Sharks

Spiny sharks, more formally called "Acanthodians" (having spines), constitute the class Acanthodii. They first appeared by the late Silurian ~420 Ma, and were among the first fishes to evolve jaws. They share features with both cartilaginous fish and bony fish, but they are not true sharks, though leading to them. They became extinct before the end of the Permian ~250 Ma. However, scales and teeth attributed to this group, as well as more derived jawed fish, such as cartilaginous and bony fish, date from the Ordovician ~460 Ma. Acanthodians were generally small shark-like fishes varying from toothless filter-feeders to toothed predators. They were once often classified as an order of the class Placodermi, but recent authorities tend to place the acanthodians as a paraphyletic assemblage leading to modern cartilaginous fish. They are distinguished in two respects: they were the earliest known jawed vertebrates, and they had stout spines supporting all their fins, fixed in place and non-movable (like a shark's dorsal fin), an important defensive adaptation. Their fossils are extremely rare.

Spiny sharks

Placoderms

Placoderms, (plate-like skin), are a group of armoured jawed fishes, of the class Placodermi. The oldest fossils appeared during the late Silurian, and became extinct at the end of the Devonian. Recent studies suggest that the placoderms are possibly a paraphyletic group of basal jawed fishes, and the closest relatives of all living jawed vertebrates. Some placoderms were small, flattened bottom-dwellers, such as antiarchs. However many, particularly the arthrodires, were active midwater predators. Dunkleosteus, which appeared later in the Devonian below, was the largest and most famous of these. The upper jaw was firmly fused to the skull, but there was a hinge joint between the skull and the bony plating of the trunk region. This allowed the upper part of the head to be thrown back and, in arthrodires, allowed them to take larger bites.

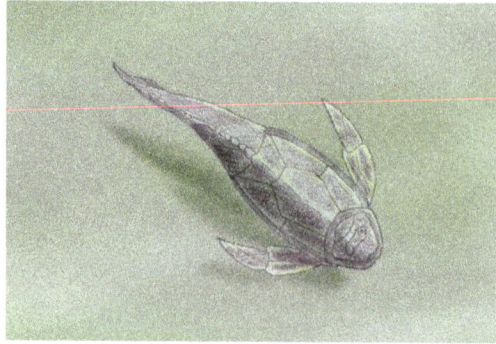
Placoderms

Guiyu Oneiros

Guiyu oneiros, the earliest known bony fish. It has the combination of both ray-finned and lobe-finned features, although analysis of the totality of its features place it closer to lobe-finned fish.

Guiyu oneiros

Andreolepis

The extinct genus Andreolepis includes the earliest known ray finned fish Andreolepis hedei, which appeared in the late Silurian, around 420 Ma.

Devonian: Age of Fishes

The Devonian Period is broken into the Early, Middle and Late Devonian. By the start of the Early Devonian 419 mya, jawed fishes had divided into four distinct clades: the placoderms and spiny sharks, both of which are now extinct, and the cartilaginous and bony fishes, both of which are still extant. The modern bony fishes, class Osteichthyes, appeared in the late Silurian or early Devonian, about 416 million years ago. Both the cartilaginous and bony fishes may have arisen from either the placoderms or the spiny sharks. A subclass of bony fishes, the ray-finned fishes (Actinopterygii), have become the dominant group in the post-Paleozoic and modern world, with some 30,000 living species.

Sea levels in the Devonian were generally high. Marine faunas were dominated by bryozoa, diverse and abundant brachiopods, the enigmatic hederelloids, microconchids and corals. Lily-like crinoids were abundant, and trilobites were still fairly common. Among vertebrates, jawless armoured fish (ostracoderms) declined in diversity, while the jawed fish (gnathostomes) simultaneously increased in both the sea and fresh water. Armoured placoderms were numerous during the lower stages of the Devonian Period but became extinct in the Late Devonian, perhaps because of competition for food against the other fish species. Early cartilaginous (Chondrichthyes) and bony fishes (Osteichthyes) also become diverse and played a large role within the Devonian seas. The first abundant genus of shark, Cladoselache, appeared in the oceans during the Devonian Period. The great diversity of fish around at the time have led to the Devonian being given the name "The Age of Fish" in popular culture.

The first ray-finned and lobe-finned bony fish appeared in the Devonian, while the placoderms began dominating almost every known aquatic environment. However, another subclass of Osteichthyes, the Sarcopterygii, including lobe-finned fishes including coelacanths and lungfish) and tetrapods, was the most diverse group of bony fishes in the Devonian. Sarcopterygians are basally characterized by internal nostrils, lobe fins containing a robust internal skeleton, and cosmoid scales.

During the Middle Devonian 393–383 Ma, the armoured jawless ostracoderm fishes were declining in diversity; the jawed fish were thriving and increasing in diversity in both the oceans and freshwater. The shallow, warm, oxygen-depleted waters of Devonian inland lakes, surrounded by primitive plants, provided the environment necessary for certain early fish to develop essential characteristics such as well developed lungs and the ability to crawl out of the water and onto the land for short periods of time. Cartilaginous fishes, class Chondrichthyes, consisting of sharks, rays and chimaeras, appeared by about 395 million years ago, in the middle Devonian.

During the Late Devonian the first forests were taking shape on land. The first tetrapods appear in the fossil record over a period, the beginning and end of which are marked with extinction events. This lasted until the end of the Devonian 359 mya. The ancestors of all tetrapods began adapting to walking on land, their strong pectoral and pelvic fins gradually evolved into legs (see Tiktaalik). In the oceans, primitive sharks became more numerous than in the Silurian and the late Ordovician. The first ammonite mollusks appeared. Trilobites, the mollusk-like brachiopods and the great coral reefs, were still common.

The Late Devonian extinction occurred at the beginning of the last phase of the Devonian period, the Famennian faunal stage, (the Frasnian-Famennian boundary), about 372.2 Ma. Many fossil agnathan fishes, save for the psammosteid heterostracans, make their last appearance shortly before this event. The Late Devonian extinction crisis primarily affected the marine community, and selectively affected shallow warm-water organisms rather than cool-water organisms. The most important group affected by this extinction event were the reef-builders of the great Devonian reef-systems.

A second extinction pulse, the Hangenberg event closed the Devonian period and had a dramatic impact on vertebrate faunas. Placoderms mostly became extinct during this event, as did most members of other groups including lobe-finned fishes, acanthodians and early tetrapods in both marine and terrestrial habitats, leaving only a handful of survivors. This event has been related to glaciation in the temperate and polar zones as well as euxinia and anoxia in the seas.

Early Devonian (419–393 Ma): Devonian (419–359 mya): The start of Devonian saw the first appearance of lobe-finned fish, precursors to the tetrapods (animals with four limbs). Major groups of fish evolved during this period, often referred to as the age of fishes.

Devonian Early Devonian

Psarolepis

Psarolepis (speckled scale) is a genus of extinct lobe-finned fish that lived around 397 to 418 Ma. Fossils of Psarolepis have been found mainly in South China and described by paleontologist Xiao-bo Yu in 1998. It is not known for certain which group Psarolepis belongs, but paleontologists agree that it probably is a basal genus and seems to be close to the common ancestor of lobe-finned and ray-finned fishes.

Psarolepis

Holoptychius

Holoptychius is an extinct genus from the order of porolepiform lobe-finned fish, extant from 416 to 359 Ma. It was a streamlined predator about 50 centimetres (20 in) long (though it could grow up to 2.5 m), which fed on other bony fish. Its rounded scales and body form indicate that it could have swum quickly through the water to catch prey. Similar to other rhipidistians, it had fang-like teeth on its palate in addition to smaller teeth on the jaws. Its asymmetrical tail sported a caudal fin on its lower end. To compensate for the downward push caused by this fin placement, Holoptychius's pectoral fins were placed high on the body.

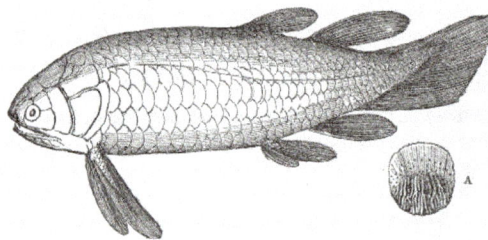

Holoptychius

Ptyctodontids

The ptyctodontids (beak-teeth) are an extinct monotypic order of unarmored placoderms, containing only one family. They were extant from the start to the end of the Devonian. With their big heads, big eyes, and long bodies, the ptyctodontids bore a strong resemblance to modern day chimaeras (Holocephali). Their armor was reduced to a pattern of small plates around the head and neck. Like the extinct and related acanthothoracids, and the living and unrelated holocephalians, most of the ptyctodontids are thought to have lived near the sea bottom and preyed on shellfish.

Ptyctodontida

Petalichthyida

The Petalichthyida was an order of small, flattened placoderms that existed from the beginning of the Devonian to the Late Devonian. They were typified by splayed fins and numerous tubercles that decorated all of the plates and scales of their armour. They reached a peak in diversity during the Early Devonian and were found throughout the world. Because they had compressed body forms, it is supposed they were bottom-dwellers that chased after or ambushed smaller fish. Their diet is not clear, as none of the fossil specimens found have preserved mouth parts.

Petalichthyida

Laccognathus

Laccognathus (pitted jaw) was a genus of amphibious lobe-finned fish that existed 398–360 Ma. They were characterized by the three large pits (fossae) on the external surface of the lower jaw, which may have had sensory functions. Laccognathus grew to 1–2 metres (3–7 ft) in length. They had very short dorsoventrally flattened heads, less than one-fifth the length of the body. The skeleton was structured so large areas of skin were stretched over solid plates of bone. This bone was composed of particularly dense fibers – so dense that exchange of oxygen through the skin was unlikely. Rather, the dense ossifications served to retain water inside the body as Laccognathus traveled on land between bodies of water.

Laccognathus

Middle Devonian

Middle Devonian (393–383 Ma): Cartilaginous fishes, consisting of sharks, rays and chimaeras, appeared about 395 Ma.

Dipterus

Dipterus (two wings) is an extinct genus of lungfish from 376–361 Ma. It was about 35 centimetres (14 in) long, mostly ate invertebrates, and had lungs, not an air bladder. Like its ancestor Dipno-rhynchus it had tooth-like plates on its palate instead of real teeth. However, unlike its modern relatives, in which the dorsal, caudal, and anal fin are fused into one, its fins were still separated. Otherwise Dipterus closely resembled modern lungfish.

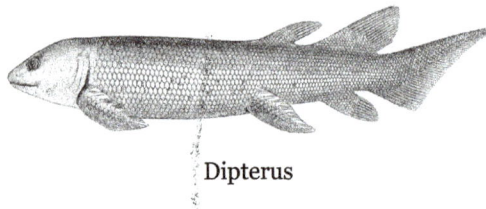
Dipterus

Cheirolepis

Cheirolepis (hand fin) was a genus of ray-finned fishes. It was among the most basal of the Devonian ray-finned fishes and is considered the first to possess the "standard" dermal cranial bones seen in later ray-finned fish. It was a predatory freshwater fish about 55 centimetres (22 in) long, and based on the size of its eyes it hunted by sight.

Cheirolepis

Cladoselache

Cladoselache was the first abundant genus of primitive shark, appearing about 370 Ma. It grew to 6 feet (1.8 m) long, with anatomical features similar to modern mackerel sharks. It had a streamlined body almost entirely devoid of scales, with five to seven gill slits and a short, rounded snout that had a terminal mouth opening at the front of the skull. It had a very weak jaw joint compared with modern-day sharks, but it compensated for that with very strong jaw-closing muscles. Its teeth were multi-cusped and smooth-edged, making them suitable for grasping, but not tearing or chewing. Cladoselache therefore probably seized prey by the tail and swallowed it whole. It had powerful keels that extended onto the side of the tail stalk and a semi-lunate tail fin, with the superior lobe about the same size as the inferior. This combination helped with its speed and agility, which was useful when trying to outswim its probable predator, the heavily armoured 10 metres (33 ft) long placoderm fish Dunkleosteus.

Cladoselache

Coccosteus

Coccosteus (seed bone) is an extinct genus of arthrodire placoderm. The majority of fossils have been found in freshwater sediments, though they may have been able to enter saltwater. They grew up to 40 centimetres (16 in) long. Like all other arthrodires, Coccosteus had a joint between the armour of the body and skull. In addition, it also had an internal joint between its neck vertebrae and the back of the skull, allowing it to open its mouth even wider. Along with the longer jaws, this allowed Coccosteus to feed on fairly large prey. As with all other arthrodires, Coccosteus had bony dental plates embedded in its jaws, forming a beak. The beak was kept sharp by having the edges of the dental plates grind away at each other.

Coccosteus

Bothriolepis

Bothriolepis (pitted scale) was the most successful genus of antiarch placoderms, if not the most successful genus of any placoderm, with over 100 species spread across Middle to Late Devonian strata across every continent.

Bothriolepis

Pituriaspida

Pituriaspida (hallucinogenic shield) is a class containing two bizarre species of armoured jawless fishes with tremendous nose-like rostrums. They lived in estuaries around 390 Ma. The paleontologist Gavin Young, named the class after the hallucinogenic drug pituri, since he thought he might be hallucinating upon viewing the bizarre forms. The better studied species looked like a throwing-dart-like, with an elongate headshield and spear-like rostrum. The other species looked like a guitar pick with a tail, with a smaller and shorter rostrum and a more triangular headshield.

Pituriaspida

Late Devonian

Late Devonian (383–359 Ma): Late Devonian extinction: 375–360 Ma. A prolonged series of extinctions eliminated about 19% of all families, 50% of all genera and 70% of all species. This extinction event lasted perhaps as long as 20 Ma, and there is evidence for a series of extinction pulses within this period.

Dunkleosteus

Dunkleosteus

Dunkleosteus is a genus of arthrodire placoderms that existed from 380 to 360 Ma. It grew up to 10 metres (33 ft) long and weighed up to 3.6 tonnes. It was a hypercarnivorous apex predator.

Apart from its contemporary Titanichthys (below), no other placoderm rivalled it in size. Instead of teeth, Dunkleosteus had two pairs of sharp bony plates, which formed a beak-like structure. Apart from megalodon, it had the most powerful bite of any fish, generating bite forces in the same league as Tyrannosaurus rex and the modern crocodile.

Titanichthys

Titanichthys is a genus of giant, aberrant marine placoderm that lived in shallow seas. Many of the species approached Dunkleosteus in size and build. Unlike its relative, however, the various species of Titanichys had small, ineffective-looking mouth-plates that lacked a sharp cutting edge. It is assumed that Titanichthys was a filter feeder that used its capacious mouth to swallow or inhale schools of small, anchovy-like fish, or possibly krill-like zooplankton, and that the mouth-plates retained the prey while allowing the water to escape as it closed its mouth.

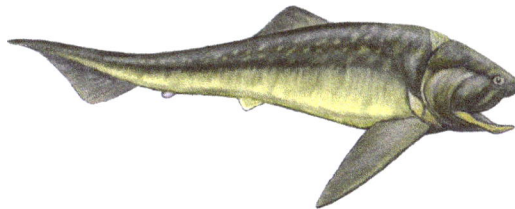

Titanichthys

Materpiscis

Materpiscis (mother fish) is a genus of ptyctodontid placoderm from about 380 Ma. Known from only one specimen, it is unique in having an unborn embryo present inside, and with remarkable preservation of a mineralised placental feeding structure (umbilical cord). This makes Materpiscis the first known vertebrate to show viviparity, or giving birth to live young. The specimen was named Materpiscis attenboroughi in honour of David Attenborough.

Materpiscis

Rhizodonts

Rhizodonts were an order of lobe-finned fish that survived to the end of the Carboniferous, 377–310 Ma. They reached huge sizes. The largest known species, Rhizodus hibberti grew up to 7 metres in length, making it the largest freshwater fish known.

Fish to Tetrapods

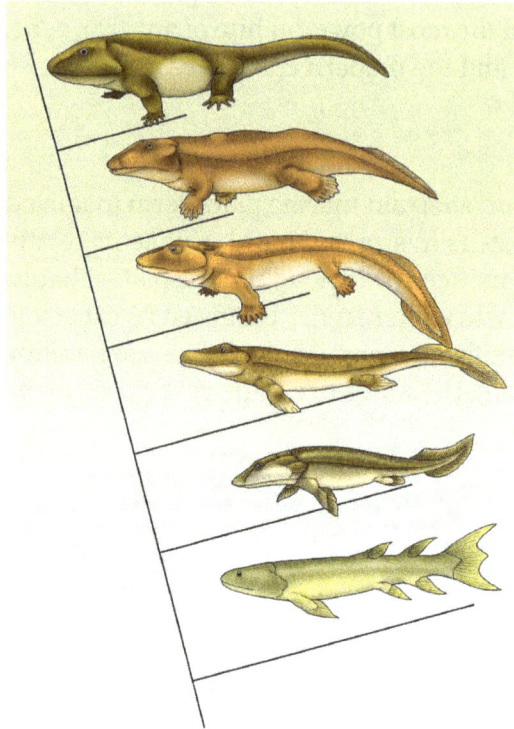

A cladogram of the evolution of tetrapods showing some of the best-known transitional fossils. It starts with Eusthenopteron at the bottom, indisputably still a fish, through Panderichthys, Tiktaalik, Acanthostega and Ichthyostega to Pederpes at the top, indisputably a tetrapod

The first tetrapods are four-legged, air-breathing, terrestrial animals from which the land vertebrates descended, including humans. They evolved from lobe-finned fish, appearing in coastal water in the middle Devonian, and giving rise to the first amphibians.

The group of lobe-finned fishes that were the ancestors of the tetrapod are grouped together as the Rhipidistia, and the first tetrapods evolved from these fish over the relatively short timespan 385–360 Ma. The early tetrapod groups themselves are grouped as Labyrinthodontia. They retained aquatic, fry-like tadpoles, a system still seen in modern amphibians. From the 1950s to the early 1980s it was thought that tetrapods evolved from fish that had already acquired the ability to crawl on land, possibly so they could go from a pool that was drying out to one that was deeper. However, in 1987, nearly complete fossils of Acanthostega from about 363 Ma showed that this Late Devonian transitional animal had legs and both lungs and gills, but could never have survived on land: its limbs and its wrist and ankle joints were too weak to bear its weight; its ribs were too short to prevent its lungs from being squeezed flat by its weight; its fish-like tail fin would have been damaged by dragging on the ground. The current hypothesis is that Acanthostega, which was about 1 metre (3.3 ft) long, was a wholly aquatic predator that hunted in shallow water. Its skeleton differed from that of most fish, in ways that enabled it to raise its head to breathe air while its body remained submerged, including: its jaws show modifications that would have enabled it to gulp air; the bones at the back of its skull are locked together, providing strong attachment points for muscles that raised its head; the head is not joined to the shoulder girdle and it has a distinct neck.

Until the 1980s early transitional lobe-finned fishes, such as the Eusthenopteron shown here, were depicted as emerging onto land. Paleontologists now widely agree this did not happen, and they were strictly aquatic.

The Devonian proliferation of land plants may help to explain why air-breathing would have been an advantage: leaves falling into streams and rivers would have encouraged the growth of aquatic vegetation; this would have attracted grazing invertebrates and small fish that preyed on them; they would have been attractive prey but the environment was unsuitable for the big marine predatory fish; air-breathing would have been necessary because these waters would have been short of oxygen, since warm water holds less dissolved oxygen than cooler marine water and since the decomposition of vegetation would have used some of the oxygen.

There are three major hypotheses as to how tetrapods evolved their stubby fins (proto-limbs). The traditional explanation is the "shrinking waterhole hypothesis" or "desert hypothesis" posited by the American paleontologist Alfred Romer. He believed limbs and lungs may have evolved from the necessity of having to find new bodies of water as old waterholes dried up.

The second hypothesis is the "inter-tidal hypothesis" put forward in 2010 by a team of Polish paleontologists led by Grzegorz Niedźwiedzki. They argued that sarcopterygians may have first emerged unto land from intertidal zones rather than inland bodies of water. Their hypothesis is based on the discovery of the 395 million-year-old Zachełmie tracks in Zachełmie, Poland, the oldest ever discovered fossil evidence of tetrapods.

The third hypothesis, the "woodland hypothesis", was proposed by the American paleontologist Gregory J. Retallack in 2011. He argues that limbs may have developed in shallow bodies of water in woodlands as a means of navigating in environments filled with roots and vegetation. He based his conclusions on the evidence that transitional tetrapod fossils are consistently found in habitats that were formerly humid and wooded floodplains.

Research by Jennifer A. Clack and her colleagues showed that the very earliest tetrapods, animals similar to Acanthostega, were wholly aquatic and quite unsuited to life on land. This is in contrast to the earlier view that fish had first invaded the land — either in search of prey (like modern mudskippers) or to find water when the pond they lived in dried out — and later evolved legs, lungs, etc.

Two ideas about the homology of arms, hands and digits have existed in the past 130 years. First that digits are unique to tetrapods and second that antecedents were present in the fins of early sarcopterygian fish. Until recently it was believed that "genetic and fossil data support the hypothesis that digits are evolutionary novelties".p. 640. However new research that created a three-dimensional reconstruction of Panderichthys, a coastal fish from the Devonian period 385 million years ago, shows that these animals already had many of the homologous bones present in the forelimbs of limbed vertebrates. For example, they had radial bones similar to rudimentary fingers but positioned in the arm-like base of their fins. Thus there was in the evolution of tetrapods a shift such that the outermost part of the fins were lost and eventually replaced by early digits. This change is consistent with additional evidence from the study of actinopterygians, sharks and lungfish that the digits of tetrapods arose from pre-existing distal radials present in more primitive fish. Controversy still exists since Tiktaalik, a vertebrate often considered the missing link between fishes and land-living animals, had stubby leg-like limbs that lacked the finger-like radial bones found in the Panderichthys. The researchers of the paper commented that it "is difficult to say whether this character distribution implies that Tiktaalik is autapomorphic, that Panderichthys and tetrapods are convergent, or that Panderichthys is closer to tetrapods than Tiktaalik. At any rate, it demonstrates that the fish–tetrapod transition was accompanied by significant character incongruence in functionally important structures.".p. 638.

From the end of the Devonian to the Mid Carboniferous a 30 million year gap occurs in the fossil record. This gap, called Romer's gap, is marked by the absence of ancestral tetrapod fossils and fossils of other vertebrates that look well-adapted for life on land.

Transition from Lobe-finned Fishes to Tetrapods

Eusthenopteron

Genus of extinct lobe-finned fishes that has attained an iconic status from its close relationships to tetrapods. Early depictions of this animal show it emerging onto land, however paleontologists now widely agree that it was a strictly aquatic animal. The genus Eusthenopteron is known from several species that lived during the Late Devonian period, about 385 Ma. It was the object of intense study from the 1940s to the 1990s by the paleoichthyologist Erik Jarvik.

Eusthenopteron~385 Ma

Gogonasus

Gogonasus (snout from Gogo) was a lobe-finned fish known from 3-dimensionally preserved 380 million-year-old fossils found in the Gogo Formation. It was a small fish reaching 30–40 cm (0.98–1.31 ft) in length. Its skeleton shows several tetrapod-like features. They included the structure of its middle ear, and its fins show the precursors of the forearm bones, the radius and ulna. Researchers believe it used its forearm-like fins to dart out of the reef to catch prey. Gogonasus was first described in 1985 by John A. Long. For almost 100 years Eusthenopteron has been the role model for demonstrating stages in the evolution of lobe-finned fishes to tetrapods. Gogonasus now replaces Eusthenopteron in being a better preserved representative without any ambiguity in interpreting its anatomy.

Gogonasus

Panderichthys

Adapted to muddy shallows, and capable of some kind of shallow water or terrestrial body flexion locomotion. Had the ability to prop itself up. They had large tetrapod-like heads, and are thought to be the most crownward stem fish-tetrapod with paired fins.

Panderichthys

Tiktaalik

A fish with limb-like fins that could take it onto land. It is an example from several lines of ancient sarcopterygian fish developing adaptations to the oxygen-poor shallow-water habitats of its time, which led to the evolution of tetrapods. Paleontologists suggest that it is representative of the transition between non-tetrapod vertebrates (fish) such as Panderichthys, known from fossils 380 million years old, and early tetrapods such as Acanthostega and Ichthyostega, known from fossils about 365 million years old. Its mixture of primitive fish and derived tetrapod characteristics led one of its discoverers, Neil Shubin, to characterize Tiktaalik as a "fishapod".

Tiktaalik

Acanthostega

Acanthostega

A fish-like early labyrinthodont that occupied swamps and changed views about the early evolution of tetrapods. It had eight digits on each hand (the number of digits on the feet is unclear) linked by webbing, it lacked wrists, and was generally poorly adapted to come onto land. Subsequent discoveries revealed earlier transitional forms between Acanthostega and completely fish-like animals.

Ichthyostega

Until finds of other early tetrapods and closely related fishes in the late 20th century, Ichthyostega stood alone as the transitional fossil between fish and tetrapods, combining a fishlike tail and gills with an amphibian skull and limbs. It possessed lungs and limbs with seven digits that helped it navigate through shallow water in swamps.

Ichthyostega 374–359 Ma

Pederpes

Pederpes is the earliest known fully terrestrial tetrapod. It is included here to complete the transition of lobe-finned fishes to tetrapods, even though Pederpes is no longer a fish.

Pederpes 359–345 Ma

By the late Devonian, land plants had stabilized freshwater habitats, allowing the first wetland ecosystems to develop, with increasingly complex food webs that afforded new opportunities. Freshwater habitats were not the only places to find water filled with organic matter and choked with plants with dense vegetation near the water's edge. Swampy habitats like shallow wetlands, coastal lagoons and large brackish river deltas also existed at this time, and there is much to suggest that this is the kind of environment in which the tetrapods evolved. Early fossil tetrapods have been found in marine sediments, and because fossils of primitive tetrapods in general are found scattered all around the world, they must have spread by following the coastal lines — they could not have lived in freshwater only.

- Fossil Illuminates Evolution of Limbs from Fins Scientific American, 2 2 April 2004.

Post Devonian

- The Mesozoic Era began about 250 million years ago in the wake of the Permian-Triassic event, the largest mass extinction in Earth's history, and ended about 66 million years ago with the Cretaceous–Paleogene extinction event, another mass extinction that killed off non-avian dinosaurs, as well as other plant and animal species. It is often referred to as the Age of Reptiles because reptiles were the dominant vertebrates of the time. The Mesozoic witnessed the gradual rifting of the supercontinent Pangaea into separate landmasses. The climate alternated between warming and cooling periods; overall the Earth was hotter than it is today.

Carboniferous

Carboniferous (359–299 Ma): Sharks underwent a major evolutionary radiation during the Carboniferous. It is believed that this evolutionary radiation occurred because the decline of the placoderms at the end of the Devonian period caused many environmental niches to become unoccupied and allowed new organisms to evolve and fill these niches.

Coastal Seas during the Carboniferous c. 300 Ma:

The first 15 million years of the Carboniferous has very few terrestrial fossils. This gap in the fossil record, is called Romer's gap after the American palaentologist Alfred Romer. While it has long been debated whether the gap is a result of fossilisation or relates to an actual event, recent work indicates the gap period saw a drop in atmospheric oxygen levels, indicating some sort of ecological collapse. The gap saw the demise of the Devonian fish-like ichthyostegalian labyrinthodonts, and the rise of the more advanced temnospondyl and reptiliomorphan amphibians that so typify the Carboniferous terrestrial vertebrate fauna.

The Carboniferous seas were inhabited by many fish, mainly Elasmobranchs (sharks and their relatives). These included some, like Psammodus, with crushing pavement-like teeth adapted for grinding the shells of brachiopods, crustaceans, and other marine organisms. Other sharks had piercing teeth, such as the Symmoriida; some, the petalodonts, had peculiar cycloid cutting teeth. Most of the sharks were marine, but the Xenacanthida invaded fresh waters of the coal swamps. Among the bony fish, the Palaeonisciformes found in coastal waters also appear to have migrated to rivers. Sarcopterygian fish were also prominent, and one group, the Rhizodonts, reached very large size.

Most species of Carboniferous marine fish have been described largely from teeth, fin spines and dermal ossicles, with smaller freshwater fish preserved whole. Freshwater fish were abundant, and include the genera Ctenodus, Uronemus, Acanthodes, Cheirodus, and Gyracanthus.

Stethacanthidae

Coastal seas during the Carboniferous c. 300 Ma

As a result of the evolutionary radiation, carboniferous sharks assumed a wide variety of bizarre shapes—including sharks of the family Stethacanthidae, which possessed a flat brush-like dorsal fin with a patch of denticles on its top. Stethacanthus' unusual fin may have been used in mating rituals. Apart from the fins, Stethacanthidae resembled Falcatus (below).

Stethacanthidae

Falcatus

Falcatus is a genus of small cladodont-toothed sharks that lived 335–318 Ma. They were about 25–30 cm (10–12 in) long. They are characterised by the prominent fin spines that curved anteriorly over their heads.

Falcatus

Orodus

Orodus is another shark of the Carboniferous, a genus from the family Orodontidae that lived into the early Permian from 303 to 295 Ma. It grew to 2 m (6.6 ft) in length.

Orodus

Permian

Permian (298–252 Ma):

Acanthodes

Acanthodes are an extinct genus of spiny shark. It had gills but no teeth, and was presumably a filter feeder. Acanthodes had only two skull bones and were covered in cubical scales. Each paired pectoral and pelvic fins had one spine, as did the single anal and dorsal fins, giving it a total of six spines, less than half that of many other spiny sharks. Acanthodians share qualities of both bony fish (osteichthyes) and cartilaginous fish (chondrichthyes), and it has been suggested that they may have been stem chondrichthyans and stem gnathostomes.

Acanthodes

The Permian ended with the most extensive extinction event recorded in paleontology: the Permian-Triassic extinction event. 90% to 95% of marine species became extinct, as well as 70% of all land organisms. It is also the only known mass extinction of insects. Recovery from the Permian-Triassic extinction event was protracted; land ecosystems took 30M years to recover, and marine ecosystems took even longer.

Triassic

Triassic (252–201 Ma): The fish fauna of the Triassic was remarkably uniform, reflecting the fact that very few families survived the Permian extinction. A considerable radiation of ray-finned fishes occurred during the Triassic, laying the foundation for many modern fishes. See Category:Triassic fish.

Perleidus

Perleidus

Perleidus was a ray-finned fish from the Early Triassic. About 15 centimetres (5.9 in) in length, it was a freshwater predatory fish with jaws that hung vertically under the braincase, allowing them to open wide. Perleidus had highly flexible dorsal and anal fins, with a reduced number of fin rays, which would have made the fish more agile in the water.

Pachycormiformes

Pachycormiformes are an extinct order of ray-finned fish that existed from the Middle Triassic to the K-Pg extinction (below). They were characterized by serrated pectoral fins, reduced pelvic fins and a bony rostrum. Their relations with other fish are unclear.

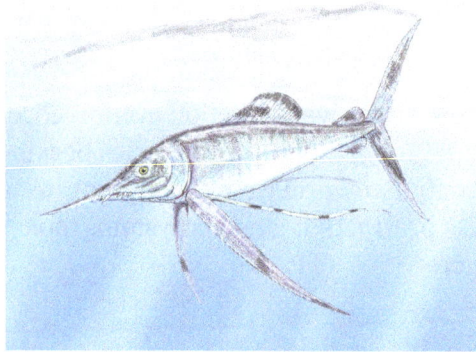

Pachycormiformes

Pholidophorus

Pholidophorus was an extinct genus of teleost, around 40 centimetres (16 in) long, from about 240–140 Ma. Although not closely related to the modern herring, it was somewhat like them. It had a single dorsal fin, a symmetrical tail, and an anal fin placed towards the rear of the body. It had large eyes and was probably a fast swimming predator, hunting planktonic crustaceans and smaller fish. A very early teleost, Pholidophoris had many primitive characteristics such as ganoid scales and a spine that was partially composed of cartilage, rather than bone.

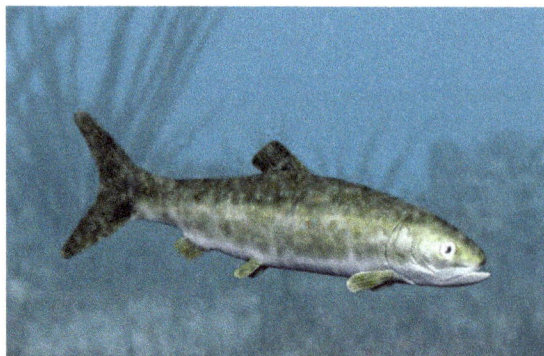

Pholidophorus

The Triassic ended with the Triassic–Jurassic extinction event. About 23% of all families, 48% of all genera (20% of marine families and 55% of marine genera) and 70% to 75% of all species became extinct. Non-dinosaurian archosaurs continued to dominate aquatic environments, while non-archosaurian diapsids continued to dominate marine environments.

Jurassic

Jurassic (201–145 Ma): During the Jurassic period, the primary vertebrates living in the seas were fish and marine reptiles. The latter include ichthyosaurs who were at the peak of their diversity, plesiosaurs, pliosaurs, and marine crocodiles of the families Teleosauridae and Metriorhynchidae. Numerous turtles could be found in lakes and rivers.

Leedsichthys

Along with its close pachycormid relatives Bonnerichthys and Rhinconichthys, Leedsichthys is part of a lineage of large-sized filter-feeders that swam the Mesozoic seas for over 100 million years, from the middle Jurassic until the end of the Cretaceous period. Pachycormids might represent an early branch of Teleostei, the group most modern bony fishes belong to; in that case Leedsichthys is the largest known teleost fish. In 2003, a fossil specimen 22 meters (72 feet) long was unearthed.

Leedsichthys

Ichthyodectidae

This fossil Ichthyodectidae from the Lower Jurassic is one of the best conserved fossil fishes worldwide

The family Ichthyodectidae (literally "fish-biters") was a family of marine actinopterygian fish. They first appeared 156 Ma during the Late Jurassic and disappeared during the K-Pg extinction event 66 Ma. They were most diverse throughout the Cretaceous period. Sometimes classified in the primitive bony fish order Pachycormiformes, they are today generally regarded as members of the "bulldog fish" order Ichthyodectiformes in the far more advanced Osteoglossomorpha. Most ichthyodectids ranged between 1 and 5 meters (3.3 and 16.4 ft) in length. All known taxa were predators, feeding on smaller fish; in several cases, larger Ichthyodectidae preyed on smaller members of the family. Some species had remarkably large teeth, though others, such as Gillicus arcuatus, had small ones and sucked in their prey. The largest Xiphactinus was 20 feet long, and appeared in the Late Cretaceous (below).

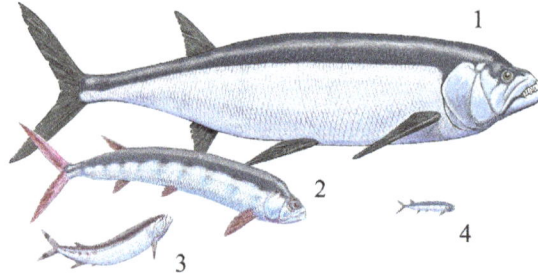

Ichthyodectidae

Cretaceous

Cretaceous (145–66 Ma):

Sturgeons

True sturgeons appear in the fossil record during the Upper Cretaceous. Since that time, sturgeons have undergone remarkably little morphological change, indicating their evolution has been exceptionally slow and earning them informal status as living fossils. This is explained in part by the long generation interval, tolerance for wide ranges of temperature and salinity, lack of predators due to size, and the abundance of prey items in the benthic environment.

Sturgeon

Cretoxyrhina

Cretoxyrhina mantelli was a large shark that lived about 100 to 82 million years ago, during the mid Cretaceous period. It is commonly known as the Ginsu Shark. This shark was first identified by a famous Swiss Naturalist, Louis Agassiz in 1843, as Cretoxyhrina mantelli. However, the most complete specimen of this shark was discovered in 1890, by the fossil hunter Charles H. Sternberg, who published his findings in 1907. The specimen consisted of a nearly complete associated vertebral column and over 250 associated teeth. This kind of exceptional preservation of fossil sharks is rare because a shark's skeleton is made of cartilage, which is not prone to fossilization. Charles dubbed the specimen Oxyrhina mantelli. This specimen represented a 20-foot-long (6.1 m) shark.

Cretoxyrhina

Enchodus

Enchodus is an extinct genus of bony fish. It flourished during the Upper Cretaceous and was small to medium in size. One of the genus' most notable attributes are the large "fangs" at the front of the upper and lower jaws and on the palatine bones, leading to its misleading nickname among fossil hunters and paleoichthyologists, "the saber-toothed herring". These fangs, along with a long sleek body and large eyes, suggest Enchodus was a predatory species.

Enchodus

Xiphactinus

Xiphactinus is an extinct genus of large predatory marine bony fish of the Late Cretaceous. They grew more than 4.5 metres (15 feet) long.

Xiphactinus

Ptychodus

Ptychodus is a genus of extinct hybodontiform shark that lived from the late Cretaceous to the Paleogene. Ptychodus mortoni (pictured) was about 32 feet (9.8 metres) long and was unearthed in Kansas, United States.

Ptychodus

The end of the Cretaceous was marked by the Cretaceous–Paleogene extinction event (K-Pg extinction). There are substantial fossil records of jawed fishes across the K–T boundary, which provides good evidence of extinction patterns of these classes of marine vertebrates. Within cartilaginous fish, approximately 80% of the sharks, rays, and skates families survived the extinction event, and more than 90% of teleost fish (bony fish) families survived. There is evidence of a mass kill of bony fishes at a fossil site immediately above the K–T boundary layer on Seymour Island near Antarctica, apparently precipitated by the K–Pg extinction event. However, the marine and freshwater environments of fishes mitigated environmental effects of the extinction event, and evidence shows that there was a major increase in size and abundance of teleosts immediately after the extinction, apparently due to the elimination of their ammonite competitors (there was no similar change in shark populations across the boundary).

Cenozoic Era

Cenozoic Era (66 Ma to present): The current era has seen great diversification of bony fishes. Over half of all living vertebrate species (about 32,000 species) are fishes (non-tetrapod craniates), a diverse set of lineages that inhabit all the world's aquatic ecosystems, from snow minnows (Cypriniformes) in Himalayan lakes at elevations over 4,600 metres (15,100 feet) to flatfishes (order Pleuronectiformes) in the Challenger Deep, the deepest ocean trench at about 11,000 metres (36,000 feet). Fishes of myriad varieties are the main predators in most of the world's water bodies, both freshwater and marine.

Amphistium

Amphistium is a 50-million-year-old fossil fish that has been identified as an early relative of the flatfish, and as a transitional fossil. In a typical modern flatfish, the head is asymmetric with both eyes on one side of the head. In Amphistium, the transition from the typical symmetric head of a vertebrate is incomplete, with one eye placed near the top of the head.

Amphistium

Megalodon

Megalodon is an extinct species of shark that lived about 28 to 1.5 Ma. It looked much like a stocky version of the great white shark, but was much larger with fossil lengths reaching 20.3 metres (67 ft). Found in all oceans it was one of the largest and most powerful predators in vertebrate history, and probably had a profound impact on marine life.

Megalodon

Prehistoric fish

Prehistoric fish are early fish that are known only from fossil records. They are the earliest known vertebrates, and include the first and extinct fish that lived through the Cambrian to the Tertiary. The study of prehistoric fish is called paleoichthyology. A few living forms, such as the coelacanth are also referred to as prehistoric fish, or even living fossils, due to their current rarity and similarity to extinct forms. Fish that have become recently extinct are not usually referred to as prehistoric fish.

Living Fossils

The jawless hagfish is a living fossil that essentially has not changed for 300 million years.

Bony Fishes

- Arowana and Arapaima
- Bowfin
- Coelacanth
- Gar
- Queensland lungfish
- Sturgeons and paddlefish
- Bichir
- Polypterus retropinnis

Sharks

- Blind shark
- Bullhead shark
- Elephant shark
- Frilled shark
- Goblin shark
- Gulper shark

Jawless Fishes

- Hagfish
- Northern brook lamprey

Eels

- Protoanguilla palau

The coelacanth was thought to have gone extinct 66 million years ago, until a living specimen belonging to the order was discovered in 1938.

Fossil sites

Miguasha National Park: outcrop of Devonian beds rich in fossil fish

Some fossil sites that have produced notable fish fossils

- Abbey Wood SSSI
- Bracklesham Beds
- Bear Gulch Limestone
- Burgess Shale
- Canowindra
- Crato Formation
- Dura Den
- Feltville Formation
- Fossil Butte National Monument
- Fur Formation
- Gogo Formation
- Green's Creek
- Green River Formation
- Kakwa Provincial Park
- Land Grove Quarry
- Maotianshan Shales
- Matanuska Formation

- McAbee Fossil Beds
- Miguasha National Park
- MoClay
- Monte Bolca
- Mount Ritchie
- Orcadian Basin
- Portishead Pier to Black Nore SSSI
- Santana Formation
- Southerham Grey Pit
- Thanet Formation
- Towaco Formation
- Weydale
- Zhoukoudian

Fossil collections

Some notable fossil fish collections.

- Fossil fish collection Natural History Museum, Britain.
- Collection and expertise Museum für Naturkunde, Germany.
- Fossil fishes The Field Museum, United States.

Paleoichthyologists

Paleoichthyology is the scientific study of the prehistoric life of fish. Listed below are some researchers who have made notable contributions to paleoichthyology.

- Louis Agassiz
- Mary Anning
- Michael Benton
- Derek Briggs
- Hans C. Bjerring
- John Samuel Budgett
- Frederick Chapman
- Jenny Clack

- Ted Daeschler
- Bashford Dean
- Robert Dick
- Philip Grey Egerton
- Edwin Sherbon Hills
- Jeffrey A. Hutchings
- Thomas Henry Huxley
- Johan Aschehoug Kiær

- Philippe Janvier
- Erik Jarvik
- George V. Lauder
- John A. Long
- Hugh Miller
- Charles Moore
- Paul E. Olsen
- Heinz Christian Pander
- Elizabeth Philpot
- Jean Piveteau

- Colin Patterson
- Alfred Romer
- Ira Rubinoff
- Neil Shubin
- Franz Steindachner
- Erik Stensiö
- Ramsay Heatley Traquair
- Thomas Stanley Westoll
- Tiberius Cornelis Winkler
- Arthur Smith Woodward

References

- Donoghue, P.C.J.; Forey, P.L.; Aldridge, R.J. (2000). "Conodont affinity and chordate phylogeny". Biological Reviews. 75 (2): 191–251. PMID 10881388. doi:10.1017/S0006323199005472. Retrieved 2008-04-07

- Gewin, V (2005). "Functional genomics thickens the biological plot". PLOS Biology. 3 (6): e219. PMC 1149496. PMID 15941356. doi:10.1371/journal.pbio.0030219

- Purnell, M. A. (2001). Derek E. G. Briggs and Peter R. Crowther, ed. Palaeobiology II. Oxford: Blackwell Publishing. p. 401. ISBN 0-632-05149-3

- Briggs, D. (May 1992). "Conodonts: a major extinct group added to the vertebrates". Science. 256 (5061): 1285–1286. Bibcode:1992Sci...256.1285B. PMID 1598571. doi:10.1126/science.1598571

- Dell'Amore, C. (September 12, 2011). "Ancient Toothy Fish Found in Arctic—Giant Prowled Rivers". National Geographic Daily News. Retrieved September 13, 2011

- Zhao Wen-Jin; Zhu Min (2007). "Diversification and faunal shift of Siluro-Devonian vertebrates of China". Geological Journal. 42 (3–4): 351–369. doi:10.1002/gj.1072

- Milsom, Clare; Rigby, Sue (2004). "Vertebrates". Fossils at a Glance. Victoria, Australia: Blackwell Publishing. p. 88. ISBN 0-632-06047-6

- Shubin N, Tabin C, Carroll S (1997). "Fossils, genes and the evolution of animal limbs". Nature. 388 (6643): 639–48. PMID 9262397. doi:10.1038/41710

- Tom Avril (September 12, 2011). "Fish fossil sheds light on 'Euramerica' phase". The Inquirer. Retrieved September 15, 2011

- Sansom, R. S. (2009). "Phylogeny, classification and character polarity of the Osteostraci (Vertebrata)". Journal of Systematic Palaeontology. 7: 95–11. doi:10.1017/S1477201908002551

- Palmer, D., ed. (1999). The Marshall Illustrated Encyclopedia of Dinosaurs and Prehistoric Animals. London: Marshall Editions. p. 23. ISBN 1-84028-152-9

- Matt Friedman (2008-07-10). "The evolutionary origin of flatfish asymmetry". Nature. 454 (7201): 209–212. Bibcode:2008Natur.454..209F. PMID 18615083. doi:10.1038/nature07108

- "Fish-Tetrapod Transition Got A New Hypothesis In 2011". Science 2.0. December 27, 2011. Retrieved January 2, 2012

- Zinsmeister WJ (1 May 1998). "Discovery of fish mortality horizon at the K–T boundary on Seymour Island: Re-evaluation of events at the end of the Cretaceous". Journal of Paleontology. 72 (3): 556–571. Retrieved 2007-08-27

- Rafferty, John P (2010) The Mesozoic Era: Age of Dinosaurs Page 219, Rosen Publishing Group. ISBN 9781615301935

- Shanta Barley (6 January 2010). "Oldest footprints of a four-legged vertebrate discovered". New Scientist. Retrieved January 3, 2010

- Shu, D-G.; et al. (November 4, 1999). "Lower Cambrian vertebrates from south China". Nature. 402 (6757): 42–46. Bibcode:1999Natur.402...42S. doi:10.1038/46965

- Sansom, Robert S. (2009). "Phylogeny, classification, & character polarity of the Osteostraci (Vertebrata)". Journal of Systematic Palaeontology. 7: 95–115. doi:10.1017/S1477201908002551

- Patterson, C (1993). Osteichthyes: Teleostei. In: The Fossil Record 2 (Benton, MJ, editor). Springer. pp. 621–656. ISBN 0-412-39380-8

- Robertson DS, McKenna MC, Toon OB, Hope S, Lillegraven JA (2004). "Survival in the first hours of the Cenozoic" (PDF). GSA Bulletin. 116 (5–6): 760–768. Bibcode:2004GSAB..116..760R. doi:10.1130/B25402.1. Retrieved 2016-01-06

Classification of Fishes

Fishes can be classified into osteichthyes, chondrichthyes and agnatha. Chondrichthyes is a class of fish which can be further divided into elasmobranchii and holocephali. The topics discussed in the chapter are of great importance to broaden the existing knowledge on fishes.

Osteichthyes

Osteichthyes is a diverse taxonomic group of fish that have skeletons primarily composed of bone tissue, as opposed to cartilage. The vast majority of fish are members of Osteichthyes, which is an extremely diverse and abundant group consisting of 45 orders, and over 435 families and 28,000 species. It is the largest class of vertebrates in existence today. The group Osteichthyes is divided into the ray-finned fish (Actinopterygii) and lobe-finned fish (Sarcopterygii). The oldest known fossils of bony fish are about 420 million years ago, which are also transitional fossils, showing a tooth pattern that is in between the tooth rows of sharks and bony fishes.

Osteichthyes can be compared to Euteleostomi. In paleontology, the terms are synonymous. In ichthyology, the difference is that Euteleostomi presents a cladistic view which includes the terrestrial tetrapods that evolved from lobe-finned fish, whereas on a traditional view, Osteichthyes includes only fishes and is therefore paraphyletic. However, recently published phylogenetic trees treat the Osteichthyes as a clade.

Characteristics

Guiyu oneiros, the earliest known bony fish, lived during the Late Silurian, 419 million years ago). It has the combination of both ray-finned and lobe-finned features, although analysis of the totality of its features place it closer to lobe-finned fish.

Bony fish are characterized by a relatively stable pattern of cranial bones, rooted, medial insertion of mandibular muscle in the lower jaw. The head and pectoral girdles are covered with large dermal bones. The eyeball is supported by a sclerotic ring of four small bones, but this characteristic has been lost or modified in many modern species. The labyrinth in the inner ear contains large otoliths. The braincase, or neurocranium, is frequently divided into anterior and posterior sections divided by a fissure.

Early bony fish had simple lungs (a pouch on either side of the esophagus) which helped them breathe in low-oxygen water. In many bony fish these have evolved into swim bladders, which help the body create a neutral balance between sinking and floating. (The lungs of amphibians, reptiles, birds, and mammals were inherited from their bony fish ancestors.) They do not have fin spines, but instead support the fin with lepidotrichia (bone fin rays). They also have an operculum, which helps them breathe without having to swim.

Bony fish have no placoid scales. Mucus glands coat the body. Most have smooth and overlapping ganoid, cycloid or ctenoid scales.

Classification

Traditionally, Osteichthyes is considered a class, recognised on having a swim bladder, only three pairs of gill arches, hidden behind a bony operculum and a predominately bony skeleton. Under this classification systems, the Osteichthyes are paraphyletic with regard to land vertebrates as the common ancestor of all Osteichthyes includes tetrapods amongst its descendants. The largest subclass, the Actinopterygii (ray-finned fish) are monophyletic, but with the inclusion of the smaller sub-class Sarcopterygii, Osteichthyes is paraphyletic.

This has led to an alternative classification, splitting the Osteichthyes into two full classes. Paradoxically, Sarcopterygii is under this scheme monophyletic, as it includes the tetrapods, making it a synonym of the clade Euteleostomi. Most bony fish belong to the ray-finned fish (Actinopterygii).

Actinopterygii

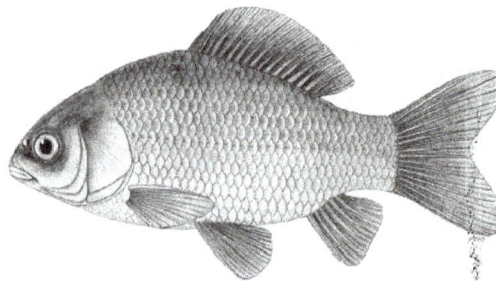

ray-finned fish

Actinopterygii, or ray-finned fishes, constitute a class or subclass of the bony fishes. The ray-finned fishes are so called because they possess lepidotrichia or "fin rays", their fins being webs of skin supported by bony or horny spines ("rays"), as opposed to the fleshy, lobed fins that characterize the class Sarcopterygii which also possess lepidotrichia. These actinopterygian fin rays attach directly to the proximal or basal skeletal elements, the radials, which represent the link or connection between these fins and the internal skeleton (e.g., pelvic and pectoral girdles). In terms of numbers, actinopterygians are the dominant class of vertebrates, comprising nearly 99% of the over 30,000 species of fish (Davis, Brian 2010). They are ubiquitous throughout freshwater and marine environments from the deep sea to the highest mountain streams. Extant species can range in size from *Paedocypris*, at 8 mm (0.3 in), to the massive ocean sunfish, at 2,300 kg (5,070 lb), and the long-bodied oarfish, to at least 11 m (36 ft).

Sarcopterygii

lobe-finned fish

Sarcopterygii (fleshy fin) or lobe-finned fish constitute a clade (traditionally a class or subclass of fish only, i.e. excluding the tetrapods) of the bony fish, though a strict cladistic view includes the terrestrial vertebrates. The living sarcopterygians are the coelacanths, lungfish, and the tetrapods. Early lobe-finned fishes had fleshy, lobed, paired fins, joined to the body by a single bone. Their fins differ from those of all other fish in that each is borne on a fleshy, lobelike, scaly stalk extending from the body. Pectoral and pelvic fins have articulations resembling those of tetrapod limbs. These fins evolved into legs of the first tetrapod land vertebrates, amphibians. They also possess two dorsal fins with separate bases, as opposed to the single dorsal fin of actinopterygians (ray-finned fish). The braincase of sarcoptergygians primitively has a hinge line, but this is lost in tetrapods and lungfish. Many early lobe-finned fishes have a symmetrical tail. All lobe-finned fishes possess teeth covered with true enamel.

Biology

All bony fish possess gills. For the majority this is their sole or main means of respiration. Lungfish and other osteichthyan species are capable of respiration through lungs or vascularized swim bladders. Other species can respire through their skin, intestines, and/or stomach.

Osteichthyes are primitively ectothermic (cold blooded), meaning that their body temperature is dependent on that of the water. But some of the larger marine osteichthyids, such as the opah, swordfish and tuna have independently evolved various levels of endothermy. Bony fish can be any type of heterotroph: numerous species of omnivore, carnivore, herbivore, filter-feeder or detritivore are documented.

Some bony fish are hermaphrodites, and a number of species exhibit parthenogenesis. Fertilization is usually external, but can be internal. Development is usually oviparous (egg-laying) but can be ovoviviparous, or viviparous. Although there is usually no parental care after birth, before birth parents may scatter, hide, guard or brood eggs, with sea horses being notable in that the males undergo a form of "pregnancy", brooding eggs deposited in a ventral pouch by a female.

Examples

The ocean sunfish is the heaviest bony fish in the world, while the longest is the king of herrings, a type of oarfish. Specimens of ocean sunfish have been observed up to 3.3 metres (11 ft) in length and weighing up to 2,303 kilograms (5,077 lb). Other very large bony fish include the Atlantic blue marlin, some specimens of which have been recorded as in excess of 820 kilograms (1,810 lb), the black marlin, some sturgeon species, and the giant and goliath grouper, which both can exceed 300 kilograms (660 lb) in weight. In contrast, the dwarf pygmy goby measures a minute 15 millimetres (0.59 in).

Arapaima gigas is the largest species of freshwater bony fish. The largest bony fish ever was *Leedsichthys*, which dwarfed the beluga sturgeon, ocean sunfish, giant grouper, and all the other giant bony fishes alive today.

Comparison with Cartilaginous Fishes

Cartilaginous fishes can be further divided into sharks, rays and chimaeras. In the table below, the comparison is made between sharks and bony fishes. For the further differences with rays, see sharks versus rays.

Comparison of cartilaginous and bony fishes		
Characteristic	**Sharks (cartilaginous)**	**Bony fishes**
Habitat	Mainly marine	Marine and freshwater
Shape	Usually dorso-ventrally flattened	Usually bilaterally flattened
Exoskeleton	Separate dermal placoid scales	Overlapping dermal cosmoid, ganoid, cycloid or ctenoid scales
Endoskeleton	Cartilaginous	Mostly bony
Caudal fin	Heterocercal	Heterocercal or diphycercal
Pelvic fins	Usually posterior.	Mostly anterior, occasionally posterior.
Intromittent organ	Males use pelvic fins as claspers for transferring sperm to a female	Do not use claspers, though some species use their anal fins as gonopodium for the same purpose
Mouth	Large, crescent shaped on the ventral side of the head	Variable shape and size at the tip or terminal part of the head
Jaw suspension	Hyostylic	Hyostylic and autostylic
Gill openings	Usually five pairs of gill slits which are not protected by an operculum.	Five pairs of gill slits protected by an operculum (a lateral flap of skin).
Type of gills	Larnellibranch with long interbranchial septum	Filiform with reduced interbranchial septum
Spiracles	The first gill slit usually becomes spiracles opening behind the eyes.	No spiracles
Afferent branchial vessels	Five pairs from ventral aorta to gills	Only four pairs
Efferent branchial vessels	Nine pairs	Four pairs
Conus arteriosus	Present in heart	Absent
Cloaca	A true cloaca is present only in cartilaginous fishes and lobe-finned fishes.	In most bony fishes, the cloaca is absent, and the anus, urinary and genital apertures open separately
Stomach	Typically J-shaped	Shape variable. Absent in some.

Intestine	Short with spiral valve in lumen	Long with no spiral valve
Rectal gland	Present	Absent
Liver	Usually has two lobes	Usually has three lobes
Swim bladder	Absent	Usually present
Brain	Has large olfactory lobes and cerebrum with small optic lobes and cerebellum	Has small olfactory lobes and cerebrum and large optic lobes and cerebellum
Restiform bodies	Present in brain	Absent
Ductus endolymphaticus	Opens on top of head	Does not open to exterior
Retina	Lacks cones	Most fish have double cones, a pair of cone cells joined to each other.
Accommodation of eye	Accommodate for near vision by moving the lens closer to the retina	Accommodate for distance vision by moving the lens further from the retina
Ampullae of Lorenzini	Present	Absent
Male genital duct	Connects to the anterior part of the genital kidney	No connection to kidney
Oviducts	Not connected to ovaries	Connected to ovaries
Urinary and genital apertures	United and urinogenital apertures lead into common cloaca	Separate and open independently to exterior
Eggs	A small number of large eggs with plenty of yolk	A large number of small eggs with little yolk
Fertilisation	Internal	Usually external
Development	Ovoviviparous types develop internally. Oviparous types develop externally using egg cases	Normally develop externally without an egg case

Actinopterygii

Actinopterygii the ray-finned fishes, constitute a class or subclass of the bony fishes.

The ray-finned fishes are so called because they possess lepidotrichia or "fin rays", their fins being webs of skin supported by bony or horny spines ("rays"), as opposed to the fleshy, lobed fins that characterize the class Sarcopterygii which also, however, possess lepidotrichia. These actinopterygian fin rays attach directly to the proximal or basal skeletal elements, the radials, which represent the link or connection between these fins and the internal skeleton (e.g., pelvic and pectoral girdles).

Numerically, actinopterygians are the dominant class of vertebrates, comprising nearly 99% of the over 30,000 species of fish. They are ubiquitous throughout freshwater and marine environments from the deep sea to the highest mountain streams. Extant species can range in size from *Paedocypris*, at 8 mm (0.3 in), to the massive ocean sunfish, at 2,300 kg (5,070 lb), and the long-bodied oarfish, at 11 m (36 ft).

Characteristics

Ray-finned fishes occur in many variant forms. The main features of a typical ray-finned fish are shown in the diagram at the left.

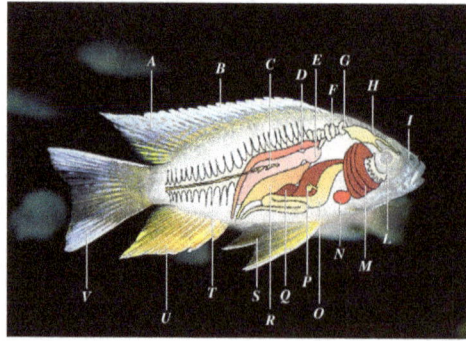

Anatomy of a typical ray-finned fish
A – dorsal fin: B – fin rays: C – lateral line: D – kidney: E – swim bladder: F – Weberian apparatus: G – inner ear:
H – brain: I – nostrils: L – eye: M – gills: N – heart O – stomach: P – gall bladder: Q – spleen: R – internal sex organs
(ovaries or testes): S – ventral fins: T – spine: U – anal fin: V – tail (caudal fin). Possible other parts not shown: barbels,
adipose fin, external genitalia (gonopodium)

Reproduction

Three-spined stickleback males (red belly) build nests and compete to attract females to
lay eggs in them. Males then defend and fan the eggs. Painting by Alexander Francis Lydon, 1879

In nearly all ray-finned fish, the sexes are separate, and in most species the females spawn eggs that are fertilized externally, typically with the male inseminating the eggs after they are laid. Development then proceeds with a free-swimming larval stage. However other patterns of ontogeny exist, with one of the commonest being sequential hermaphroditism. In most cases this involves protogyny, fish starting life as females and converting to males at some stage, triggered by some internal or external factor. This may be advantageous as females become less prolific as they age while male fecundity increases with age. Protandry, where a fish converts from male to female, is much less common than protogyny. Most families use external rather than internal fertilization. Of the oviparous teleosts, most (79%) do not provide parental care. Viviparity, ovoviviparity, or some form of parental care for eggs, whether by the male, the female, or both parents is seen in a significant fraction (21%) of the 422 teleost families; no care is likely the ancestral condition. Viviparity is relatively rare and is found in about 6% of teleost species; male care is far more common than female care. Male territoriality "preadapts" a species for evolving male parental care.

There are a few examples of fish that self-fertilise. The mangrove rivulus is an amphibious, simultaneous hermaphrodite, producing both eggs and spawn and having internal fertilisation. This mode of reproduction may be related to the fish's habit of spending long periods out of water in the mangrove forests it inhabits. Males are occasionally produced at temperatures below 19 °C (66 °F) and can fertilise eggs that are then spawned by the female. This maintains genetic variability in a species that is otherwise highly inbred.

Fossil Record

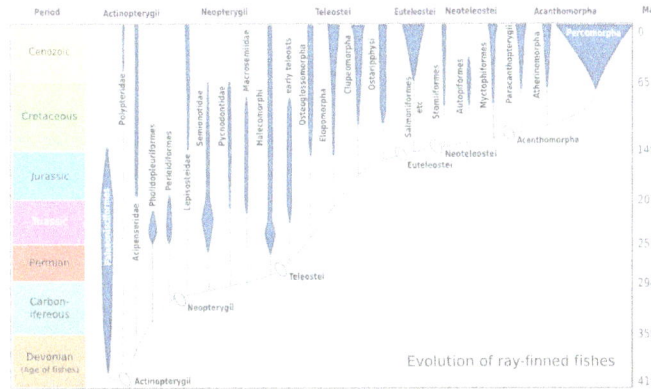

Evolution of ray-finned fishes

The earliest known fossil actinopterygiian is Andreolepis hedei, dating back 420 million years (Late Silurian). Remains have been found in Russia, Sweden, and Estonia.

Classification

Actinopterygians are divided into the subclasses Chondrostei and Neopterygii. The Neopterygii, in turn, are divided into the infraclasses Holostei and Teleostei. During the Mesozoic and Cenozoic the teleosts in particular diversified widely, and as a result, 96% of all known fish species are teleosts. The cladogram shows the major groups of actinopterygians and their relationship to the terrestrial vertebrates (tetrapods) that evolved from a related group of fish. Approximate dates are from Near et al., 2012.

The polypterids (bichirs and ropefish) are the sister lineage of all other actinopterygians, The Acipenseriformes (sturgeons and paddlefishes) are the sister lineage of Neopterygii, and Holostei (bowfin and gars) are the sister lineage of teleosts. The Elopomorpha (eels and tarpons) appears to be the most basic teleosts.

Chondrostei

Atlantic sturgeon

Chondrostei *(cartilage bone)* are primarily cartilaginous fish showing some ossification. There are 52 species divided among two orders, the Acipenseriformes (sturgeons and paddlefishes) and the Polypteriformes (reedfishes and bichirs). It is thought that the chondrosteans evolved from bony fish but lost the bony hardening of their cartilaginous skeletons, resulting in a lightening of the frame. Elderly chondrosteans show beginnings of ossification of the skeleton, suggesting that this process is delayed rather than lost in these fish. This group has at times been classified with the sharks: the similarities are obvious, as not only do the chondrosteans mostly lack bone, but the structure of the jaw is more akin to that of sharks than other bony fish, and both lack scales (excluding the Polypteriforms). Additional shared features include spiracles and, in sturgeons, a heterocercal tail (the vertebrae extend into the larger lobe of the caudal fin). However the fossil record suggests that these fish have more in common with the Teleostei than their external appearance might suggest. Chondrostei is paraphyletic meaning that this subclass does not contain all the descendants of their common ancestor; reclassification of the Chondrostei is therefore not out of the question.

Neopterygii

Atlantic salmon

Neopterygii *(new fins)* appeared somewhere in the Late Permian, before the time of the dinosaurs. There are only few changes during their evolution from the earlier actinopterygians. They are a very successful group of fishes, because they can move more rapidly than their ancestors. Their scales and skeletons began to lighten during their evolution, and their jaws became more powerful and efficient. While electroreception and the ampullae of Lorenzini is present in all other groups of fish, with the exception of hagfish, Neopterygii has lost this sense, though it later re-evolved within Gymnotiformes and catfishes, who possess nonhomologous teleost ampullae.

Sarcopterygii

The Sarcopterygii or lobe-finned fish – sometimes considered synonymous with Crossopterygii– constitute a clade (traditionally a class or subclass) of the bony fish, though a strict cladistic view includes the terrestrial vertebrates.

The living sarcopterygians are the coelacanths and lungfish; additionally, all tetrapods are sarcopterygians or descendants of them (including humans).

Characteristics

Early lobe-finned fishes are bony fish with fleshy, lobed, paired fins, which are joined to the body by a single bone. The fins of lobe-finned fishes differ from those of all other fish in that each is borne on a fleshy, lobelike, scaly stalk extending from the body. The scales of sarcopterygians are true scaloids, consisting of lamellar bone surrounded by layers of vascular bone, dentine-like cosmine, and external keratin. Pectoral and pelvic fins have articulations resembling those of tetrapod limbs. These fins evolved into the legs of the first tetrapod land vertebrates, amphibians. They also possess two dorsal fins with separate bases, as opposed to the single dorsal fin of actinopterygians (ray-finned fish). The braincase of sarcopterygians primitively has a hinge line, but this is lost in tetrapods and lungfish. Many early sarcopterygians have a symmetrical tail. All sarcopterygians possess teeth covered with true enamel.

Most species of lobe-finned fishes are extinct. The largest known lobe-finned fish was *Rhizodus hibberti* from the Carboniferous period of Scotland which may have exceeded 7 meters in length. Among the two groups of extant (living) species, the coelacanths and the lungfishes, the largest species is the West Indian Ocean coelacanth, reaching 2 m (6.5 ft) in length and weighing up 110 kg (240 lb). The largest lungfish is the African lungfish which can reach 2 m (6.6 ft) in length and weigh up to 50 kg (110 lb).

Classification

Taxonomists who subscribe to the cladistic approach include the grouping Tetrapoda within this group, which in turn consists of all species of four-limbed vertebrates. The fin-limbs of lobe-finned fishes such as the coelacanths show a strong similarity to the expected ancestral form of tetrapod limbs. The lobe-finned fishes apparently followed two different lines of development and are accordingly separated into two subclasses, the Rhipidistia (including the Dipnoi, the lungfish, and the Tetrapodomorpha which include the Tetrapoda) and the Actinistia (coelacanths).

Taxonomy

The classification below follows Benton 2004, and uses a synthesis of rank-based Linnaean taxonomy and also reflects evolutionary relationships. Benton included the Superclass Tetrapoda in the Subclass Sarcopterygii in order to reflect the direct descent of tetrapods from lobe-finned fish, despite the former being assigned a higher taxonomic rank.

Actinistia

West Indian Ocean coelacanth

Actinistia, coelacanths, are a subclass of mostly fossil lobe-finned fishes. This subclass contains the coelacanths, including the two living coelacanths, the West Indian Ocean coelacanth and the Indonesian coelacanth.

Dipnoi

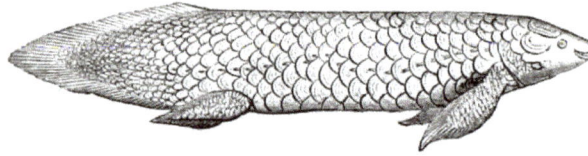

Queensland lungfish

Dipnoi, lungfish, also known as salamanderfish, are a subclass of freshwater fish. Lungfish are best known for retaining characteristics primitive within the bony fishes, including the ability to breathe air, and structures primitive within the lobe-finned fishes, including the presence of lobed fins with a well-developed internal skeleton. Today, lungfish live only in Africa, South America, and Australia. While vicariance would suggest this represents an ancient distribution limited to the Mesozoic supercontinent Gondwana, the fossil record suggests advanced lungfish had a widespread freshwater distribution and the current distribution of modern lungfish species reflects extinction of many lineages following the breakup of Pangaea, Gondwana, and Laurasia.

Tetrapodomorpha

Advanced tetrapodomorph *Tiktaalik*

Tetrapodomorpha, tetrapods and their extinct relatives, are a clade of vertebrates consisting of tetrapods (four-limbed vertebrates) and their closest sarcopterygian relatives that are more closely related to living tetrapods than to living lungfish. Advanced forms transitional between fish and the early labyrinthodonts, like *Tiktaalik*, have been referred to as "fishapods" by their discoverers, being half-fish, half-tetrapods, in appearance and limb morphology. The Tetrapodomorpha contain the crown group tetrapods (the last common ancestor of living tetrapods and all of its descendants) and several groups of early stem tetrapods, and several groups of related lobe-finned fishes, collectively known as the osteolepiforms. The Tetrapodamorpha minus the crown group Tetrapoda are the Stem Tetrapoda, a paraphyletic unit encompassing the fish to tetrapod transition. Among the characters defining tetrapodomorphs are modifications to the fins, notably a humerus with convex head articulating with the glenoid fossa (the socket of the shoulder joint). Tetrapodomorph fossils are known from the early Devonian onwards, and include *Osteolepis*, *Panderichthys*, *Kenichthys*, and *Tungsenia*.

Evolution

Evolution of Lobe-finned Fishes

Lobe-finned fishes (sarcopterygians) and their relatives the ray-finned fishes (actinopterygians) comprise the superclass of bony fishes (Osteichthyes) characterized by their bony skeleton rather than cartilage. There are otherwise vast differences in fin, respiratory, and circulatory structures between the Sarcopterygii and the Actinopterygii, such as the presence of cosmoid layers in the scales of sarcopterygians. The earliest fossils of sarcopterygians, found in the uppermost Silurian (ca 418 Ma), closely resembled the acanthodians (the "spiny fish", a taxon that became extinct at the end of the Paleozoic). In the early–middle Devonian (416 - 385 Ma), while the predatory placoderms dominated the seas, some sarcopterygians came into freshwater habitats.

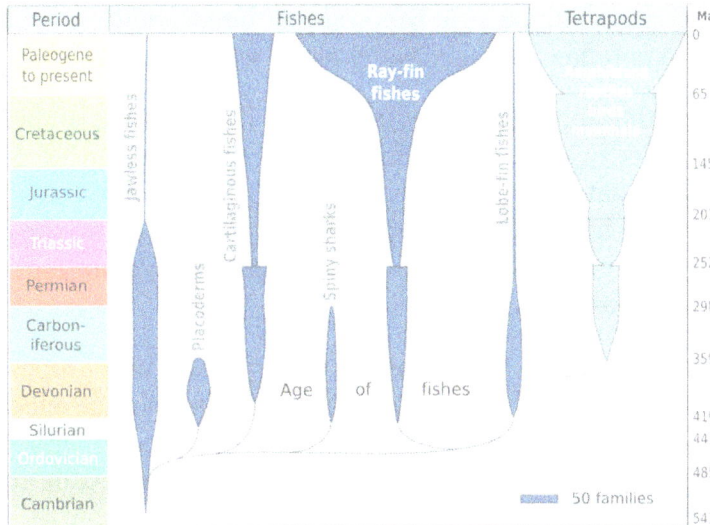

Spindle diagram for the evolution of lobe-finned fishes, tetrapods and other vertebrate classes.

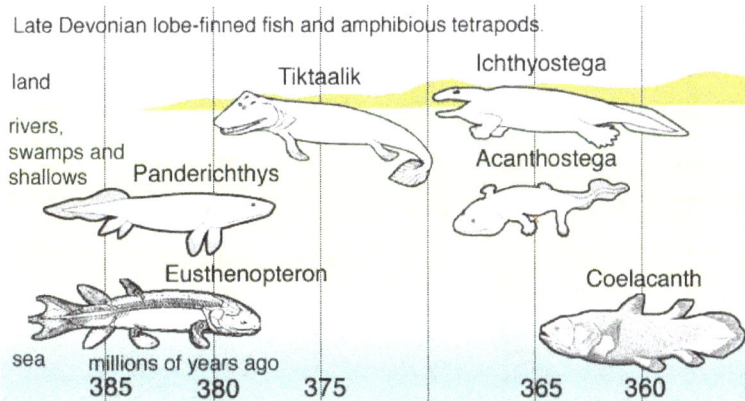

In Late Devonian vertebrate speciation, descendants of pelagic lobe-finned fish — like *Eusthenopteron* — exhibited a sequence of adaptations:
- Panderichthys, suited to muddy shallows;
- Tiktaalik with limb-like fins that could take it onto land;
- Early tetrapods in weed-filled swamps, such as:
  - ◇ Acanthostega, which had feet with eight digits,
  - ◇ Ichthyostega with limbs.

Descendants also included pelagiclobe-finned fish such as coelacanth species.

In the Early Devonian (416 - 397 Mya), the sarcopterygians split into two main lineages — the coelacanths and the rhipidistians. Coelacanths never left the oceans and their heyday was the late Devonian and Carboniferous, from 385 to 299 Ma, as they were more common during those periods than in any other period in the Phanerozoic; coelacanths still live today in the oceans (genus *Latimeria*).

The Rhipidistians, whose ancestors probably lived in the oceans near the river mouths (estuaries), left the ocean world and migrated into freshwater habitats. They in turn split into two major groups: the lungfish and the tetrapodomorphs. The lungfish's greatest diversity was in the Triassic period; today there are fewer than a dozen genera left. The lungfish evolved the first proto-lungs and proto-limbs; developing the ability to live outside a water environment in the middle Devonian (397 - 385 Ma).

There are three major hypotheses as to how they evolved their stubby fins (proto-limbs). The traditional explanation is the "shrinking waterhole hypothesis" or "desert hypothesis" posited by the American paleontologist Alfred Romer. He believed that limbs and lungs may have evolved from the necessity of having to find new bodies of water as old waterholes dried up.

The second hypothesis is the "inter-tidal hypothesis" put forward in 2010 by a team of Polish paleontologists (Grzegorz Niedźwiedzki, Piotr Szrek, Katarzyna Narkiewicz, Marek Narkiewicz, & Per Ahlberg). They argued that sarcopterygians may have first emerged unto land from intertidal zones rather than inland bodies of water. Their hypothesis is based on the discovery of the 395 million-year-old Zachełmie tracks in Zachełmie, Świętokrzyskie Voivodeship, Poland, the oldest ever discovered fossil evidence of tetrapods.

The third hypothesis is dubbed the "woodland hypothesis" and was proposed by the American paleontologist Greg Retallack in 2011. He argues that limbs may have developed in shallow bodies of water in woodlands as a means of navigating in environments filled with roots and vegetation. He based his conclusions on the evidence that transitional tetrapod fossils are consistently found in habitats that were formerly humid and wooded floodplains.

The first tetrapodomorphs, which included the gigantic rhizodonts, had the same general anatomy as the lungfish, who were their closest kin, but they appear not to have left their water habitat until the late Devonian epoch (385 - 359 Ma), with the appearance of tetrapods (four-legged vertebrates). Tetrapods are the only tetrapodomorphs which survived after the Devonian.

Non-tetrapod sarcopterygians continued until towards the end of Paleozoic era, suffering heavy losses during the Permian-Triassic extinction event (251 Ma).

Chondrichthyes

Chondrichthyes is a class that contains the cartilaginous fishes: they are jawed vertebrates with paired fins, paired nares, scales, a heart with its chambers in series, and skeletons made of cartilage rather than bone. The class is divided into two subclasses: Elasmobranchii (sharks, rays, skates, and sawfish) and Holocephali (chimaeras, sometimes called ghost sharks, which are sometimes separated into their own class).

Within the infraphylum Gnathostomata, cartilaginous fishes are distinct from all other jawed vertebrates.

Anatomy

Skeleton

The skeleton is cartilaginous. The notochord, which is present in the young, is gradually replaced by cartilage. Chondrichthyans also lack ribs, so if they leave water, the larger species' own body weight would crush their internal organs long before they would suffocate.

As they do not have bone marrow, red blood cells are produced in the spleen and the epigonal organ (special tissue around the gonads, which is also thought to play a role in the immune system). They are also produced in the Leydig's organ, which is only found in certain cartilaginous fishes. The subclass Holocephali, which is a very specialized group, lacks both the Leydig's and epigonal organs.

Appendages

Apart from electric rays, which have a thick and flabby body, with soft, loose skin, chondrichthyans have tough skin covered with dermal teeth (again, Holocephali is an exception, as the teeth are lost in adults, only kept on the clasping organ seen on the caudal ventral surface of the male), also called placoid scales (or *dermal denticles*), making it feel like sandpaper. In most species, all dermal denticles are oriented in one direction, making the skin feel very smooth if rubbed in one direction and very rough if rubbed in the other.

Originally, the pectoral and pelvic girdles, which do not contain any dermal elements, did not connect. In later forms, each pair of fins became ventrally connected in the middle when scapulo-coracoid and pubioischiadic bars evolved. In rays, the pectoral fins have connected to the head and are very flexible.

One of the primary characteristics present in most sharks is the heterocercal tail, which aids in locomotion.

Body Covering

Chondrichthyans have toothlike scales called dermal denticles or placoid scales. Denticles provide two functions, protection and in most cases, streamlining. Mucous glands exist in some species as well.

It is assumed that their oral teeth evolved from dermal denticles that migrated into the mouth, but it could be the other way around as the teleost bony fish *Denticeps clupeoides* has most of its head covered by dermal teeth (as does, probably, *Atherion elymus*, another bony fish). This is most likely a secondary evolved characteristic, which means there is not necessarily a connection between the teeth and the original dermal scales.

The old placoderms did not have teeth at all, but had sharp bony plates in their mouth. Thus, it is unknown whether the dermal or oral teeth evolved first. Nor is it sure how many times it has happened if it turns out to be the case. It has even been suggested that the original bony plates of all the vertebrates are gone and that the present scales are just modified teeth, even if both teeth and the body armor have a common origin a long time ago. However, there is no evidence of this at the moment.

Respiratory System

All chondrichthyans breathe through five to seven pairs of gills, depending on the species. In general, pelagic species must keep swimming to keep oxygenated water moving through their gills, whilst demersal species can actively pump water in through their spiracles and out through their gills. However, this is only a general rule and many species differ.

A spiracle is a small hole found behind each eye. These can be tiny and circular, such as found on the nurse shark (*Ginglymostoma cirratum*), to extended and slit-like, such as found on the wobbegongs (Orectolobidae). Many larger, pelagic species, such as the mackerel sharks (Lamnidae) and the thresher sharks (Alopiidae), no longer possess them.

Immune System

Like all other jawed vertebrates, members of Chondrichthyes have an adaptive immune system.

Reproduction

Fertilization is internal. Development is usually live birth (ovoviviparous species) but can be through eggs (oviparous). Some rare species are viviparous. There is no parental care after birth; however, some chondrichthyans do guard their eggs.

Classification

The class Chondrichthyes has two subclasses: the subclass Elasmobranchii (sharks, rays, skates, and sawfish) and the subclass Holocephali (chimaeras).

Subclasses of Cartilaginous Fishes

Elasmobranchii

Elasmobranchii is a subclass that includes the sharks and the rays and skates. Members of the elasmobranchii have no swim bladders, five to seven pairs of gill clefts opening individually to the exterior, rigid dorsal fins, and small placoid scales. The teeth are in several series; the upper jaw is not fused to the cranium, and the lower jaw is articulated with the upper. The eyes have a tapetum lucidum. The inner margin of each pelvic fin in the male fish is grooved to constitute a clasper for the transmission of sperm. These fish are widely distributed in tropical and temperate waters.

Sharks and rays, skates, and sawfish

Holocephali

Holocephali *(complete-heads)* is a subclass of which the order Chimaeriformes is the only surviving group. This group includes the rat fishes (e.g., *Chimaera*), rabbit-fishes (e.g., *Hydrolagus*) and elephant-fishes (*Callorhynchus*). Today, they preserve some features of elasmobranch life in Paleaozoic times, though in other respects they are aberrant. They live close to the bottom and feed on molluscs and other invertebrates. The tail is long and thin and they move by sweeping movements of the large pectoral fins. There is an erectile spine in front of the dorsal fin, sometimes poisonous. There is no stomach (that is, the gut is simplified and the 'stomach' is merged with the intestine), and the mouth is a small aperture surrounded by lips, giving the head a parrot-like appearance.

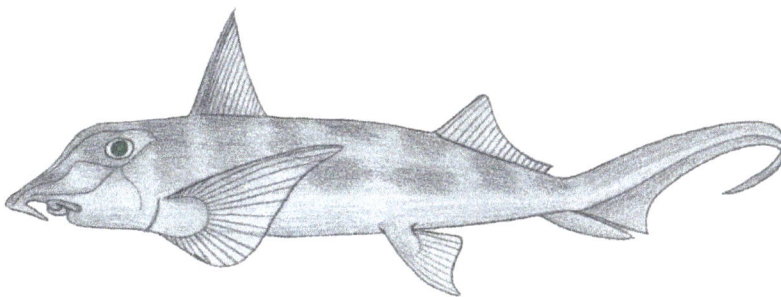

Chimaeras

The fossil record of the Holocephali starts in the Devonian period. The record is extensive, but most fossils are teeth, and the body forms of numerous species are not known, or at best poorly understood.

Extant orders of cartilaginous fishes										
Group	**Order**	**Image**	**Common name**	**Authority**	**Families**	**Genera**	**Species**			
							Total	CR	EN	VU
Galean sharks	Carcharhiniformes		ground sharks	Compagno, 1977	8	51	>270	7	10	21
	Heterodontiformes		bullhead sharks	L. S. Berg, 1940	1	1	9			
	Lamniformes		mackerel sharks	L. S. Berg, 1958	7 +2 extinct	10	16			10
	Orectolobiformes		carpet sharks	Applegate, 1972	7	13	43			7

	Order	Common name	Author						
Squalo-morph sharks	Hex-anchiformes	frilled and cow sharks	de Buen, 1926	2 +3 extinct	4 +11 extinct	6 +33 extinct			
	Pristiophori-formes	sawsharks	L. S. Berg, 1958	1	2	6			
	Squaliformes	dogfish sharks		1	2	29	1		6
	Squatiniformes	angel sharks	Buen, 1926	1	1	23	3	4	5
Rays	Myliobati-formes	stingrays and relatives	Compagno, 1973	10	29	223	1	16	33
	Pristiformes	sawfishes		1	2	5-7	5-7		
	Rajiformes	skates and guitarfishes	L. S. Berg, 1940	5	36	>270	4	12	26
	Torpedini-formes	electric rays	de Buen, 1926	2	12	69	2		9
Holo-cephali	Chimaeri-formes	chimaera	Obruchev, 1953	3 +2 extinct	6 +3 extinct	39 +17 extinct			

Evolution

Cartilaginous fish are considered to have evolved from acanthodians. Originally assumed to be closely related to bony fish or a polyphyletic assemblage leading to both groups, the discovery of *Entelognathus* and several examinations of acanthodian characteristics indicate that bony fish evolved directly from placoderm like ancestors, while acanthodians represent a paraphyletic assemblage leading to Chondrichthyes. Some characteristics previously thought to be exclusive to acanthodians are also present in basal cartilaginous fish. In particular, new phylogenetic studies find cartilaginous fish to be well nested among acanthodians, with *Doliodus* and *Tamiobatis* being the closest relatives to Chondrichthyes. Recent studies vindicate this, as *Doliodus* had a mosaic of chondrichthyian and acanthodiian traits.

Unequivocal fossils of cartilaginous fishes first appeared in the fossil record by about 395 million years ago, during the middle Devonian. The radiation of elasmobranches in the chart on the right is divided into the taxa: Cladoselache, Eugeneodontiformes, Symmoriida, Xenacanthiformes, Ctenacanthiformes, Hybodontiformes, Galeomorphi, Squaliformes and Batoidea.

By the start of the Early Devonian, 419 mya (million years ago), jawed fishes had divided into three distinct groups: the now extinct placoderms (a paraphyletic assemblage of ancient armoured fishes), the bony fishes and the clade including spiny sharks and early cartilaginous fish. The modern bony fishes, class Osteichthyes, appeared in the late Silurian or early Devonian, about 416 million years ago. Cartilaginous fishes first appeared about 395 Ma, having evolved from *Doliodus*-like spiny shark ancestors. The first abundant genus of shark, *Cladoselache*, appeared in the oceans during the Devonian Period.

A Bayesian analysis of molecular data suggests that the Holocephali and Elasmoblanchii diverged in the Silurian (421 million years ago) and that the sharks and rays/skates split in the Carboniferous (306 million years ago).

Devonian

Devonian (419–359 Mya):

Cladoselache

Cladoselache

Cladoselache was the first abundant genus of primitive shark, appearing about 370 Ma. It grew to 6 feet (1.8 m) long, with anatomical features similar to modern mackerel sharks. It had a streamlined body almost entirely devoid of scales, with five to seven gill slits and a short, rounded snout that had a terminal mouth opening at the front of the skull. It had a very weak jaw joint compared with modern-day sharks, but it compensated for that with very strong jaw-closing muscles. Its teeth were multi-cusped and smooth-edged, making them suitable for grasping, but not tearing or chewing. *Cladoselache* therefore probably seized prey by the tail and swallowed it whole. It had powerful keels that extended onto the side of the tail stalk and a semi-lunate tail fin, with the superior lobe about the same size as the inferior. This combination helped with its speed and agility which was useful when trying to outswim its probable predator, the heavily armoured 10 metres (33 ft) long placoderm fish *Dunkleosteus*.

Carboniferous

Carboniferous (359–299 Ma): Sharks underwent a major evolutionary radiation during the Carboniferous. It is believed that this evolutionary radiation occurred because the decline of the placoderms at the end of the Devonian period caused many environmental niches to become unoccupied and allowed new organisms to evolve and fill these niches.

Orthacanthus Senckenbergianus

Orthacanthus senckenbergianus

The first 15 million years of the Carboniferous has very few terrestrial fossils. This gap in the fossil record, is called Romer's gap after the American palaentologist Alfred Romer. While it has long been debated whether the gap is a result of fossilisation or relates to an actual event, recent work indicates that the gap period saw a drop in atmospheric oxygen levels, indicating some sort of ecological collapse. The gap saw the demise of the Devonian fish-like ichthyostegalian labyrinthodonts, and the rise of the more advanced temnospondyl and reptiliomorphan amphibians that so typify the Carboniferous terrestrial vertebrate fauna.

The Carboniferous seas were inhabited by many fish, mainly Elasmobranchs (sharks and their relatives). These included some, like Psammodus, with crushing pavement-like teeth adapted for grinding the shells of brachiopods, crustaceans, and other marine organisms. Other sharks had piercing teeth, such as the Symmoriida; some, the petalodonts, had peculiar cycloid cutting teeth. Most of the sharks were marine, but the Xenacanthida invaded fresh waters of the coal swamps. Among the bony fish, the Palaeonisciformes found in coastal waters also appear to have migrated to rivers. Sarcopterygian fish were also prominent, and one group, the Rhizodonts, reached very large size.

Most species of Carboniferous marine fish have been described largely from teeth, fin spines and dermal ossicles, with smaller freshwater fish preserved whole. Freshwater fish were abundant, and include the genera Ctenodus, Uronemus, Acanthodes, Cheirodus, and Gyracanthus.

Stethacanthidae

As a result of the evolutionary radiation, carboniferous sharks assumed a wide variety of bizarre shapes; e.g., sharks belonging to the family Stethacanthidae possessed a flat brush-like dorsal fin with a patch of denticles on its top. Stethacanthus' unusual fin may have been used in mating rituals. Apart from the fins, Stethacanthidae resembled Falcatus (below).

Stethacanthidae

Falcatus

Falcatus is a genus of small cladodont-toothed sharks which lived 335–318 Ma. They were about 25–30 cm (9.8–11.8 in) long. They are characterised by the prominent fin spines that curved anteriorly over their heads.

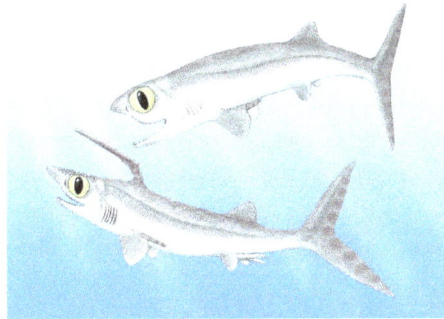

Falcatus

Orodus

Orodus is another shark of the Carboniferous, a genus from the family Orodontidae that lived into the early Permian from 303 to 295 Ma. It grew to 2 m (6.6 ft) in length.

Orodus

Permian

Permian (298–252 Ma): The Permian ended with the most extensive extinction event recorded in paleontology: the Permian-Triassic extinction event. 90% to 95% of marine species became extinct, as well as 70% of all land organisms. Recovery from the Permian-Triassic extinction event was protracted; land ecosystems took 30M years to recover, and marine ecosystems took even longer.

Triassic

Triassic (252–201 Ma): The fish fauna of the Triassic was remarkably uniform, reflecting the fact that very few families survived the Permian extinction. In turn, the Triassic ended with the Triassic–Jurassic extinction event. About 23% of all families, 48% of all genera (20% of marine families and 55% of marine genera) and 70% to 75% of all species became extinct.

Jurassic

Jurassic (201–145 Ma): The end of the Cretaceous was marked by the Cretaceous–Paleogene extinction event (K-Pg extinction). There are substantial fossil records of jawed fishes across the K–T boundary, which provides good evidence of extinction patterns of these classes of marine vertebrates. Within cartilaginous fish, approximately 80% of the sharks, rays, and skates families survived the extinction event, and more than 90% of teleost fish (bony fish) families survived.

Cretaceou

Cretaceous (145–66 Ma):

Squalicorax Falcatus

Squalicorax falcatus

Squalicorax falcatus is a lamnoid shark from the Cretaceous.

Ptychodus

Ptychodus is a genus of extinct hybodontiform shark which lived from the late Cretaceous to the Paleogene. Ptychodus mortoni (pictured) was about 32 feet (9.8 meters) long and was unearthed in Kansas, United States.

Ptychodus

Cenozoic Era

Cenozoic Era (65 Ma to present): The current era has seen great diversification of bony fishes.

Megalodon

Megalodon is an extinct species of shark that lived about 28 to 1.5 Ma. It looked much like a stocky version of the great white shark, but was much larger with fossil lengths reaching 20.3 metres (67 ft). Found in all oceans it was one of the largest and most powerful predators in vertebrate history, and probably had a profound impact on marine life.

Megalodon

Elasmobranchii

Elasmobranchs lack swim bladders, and maintain buoyancy with oil that they store in their livers. Some deep sea sharks are targeted by fisheries for this liver oil, including the school, gulper and basking sharks (pictured). All three of these species have been assessed by the IUCN as vulnerable due to overfishing.

From a practical point of view the life-history pattern of elasmobranchs makes this group of animals extremely susceptible to over fishing. It is no coincidence that the commercially exploited marine turtles and baleen whales, which have life-history patterns similar to the sharks, are also in trouble.

Elasmobranchii is a subclass of Chondrichthyes or cartilaginous fish. Members of this subclass are characterised by having five to seven pairs of gill clefts opening individually to the exterior, rigid dorsal fins and small placoid scales on the skin. The teeth are in several series; the upper jaw is not fused to the cranium, and the lower jaw is articulated with the upper. The details of this jaw anatomy vary between species, and help distinguish the different elasmobranch clades. The pelvic fins in males are modified to create claspers for the transfer of sperm. There is no swim bladder, instead these fish maintain buoyancy with large livers rich in oil.

The earliest elasmobranch fossils came from the Devonian and many surviving orders date back to the Cretaceous, or even earlier. Many species became extinct during the Permian and there was a burst of adaptive radiation during the Jurassic. Extant species are classified under Selachii (Selachimorpha), the modern sharks, and Batoidea, the rays, skates and sawfish.

Description

Elasmobranchii is one of the two subclasses of cartilaginous fish in the class Chondrichthyes, the other being Holocephali (chimaeras).

Members of the elasmobranchii subclass have no swim bladders, five to seven pairs of gill clefts opening individually to the exterior, rigid dorsal fins, and small placoid scales. The teeth are in several series; the upper jaw is not fused to the cranium, and the lower jaw is articulated with the upper.

Extant elasmobranchs exhibit several archetypal jaw suspensions: amphistyly, orbitostyly, hyostyly, and euhyostyly. In amphistyly, the palatoquadrate has a postorbital articulation with the chondrocranium from which ligaments primarily suspend it anteriorly. The hyoid articulates with the mandibular arch posteriorly, but it appears to provide little support to the upper and lower jaws. In orbitostyly, the orbital process hinges with the orbital wall and the hyoid provides the majority of suspensory support.

In contrast, hyostyly involves an ethmoid articulation between the upper jaw and the cranium, while the hyoid most likely provides vastly more jaw support compared to the anterior ligaments. Finally, in euhyostyly, also known as true hyostyly, the mandibular cartilages lack a ligamentous connection to the cranium. Instead, the hyomandibular cartilages provide the only means of jaw support, while the ceratohyal and basihyal elements articulate with the lower jaw, but are disconnected from the rest of the hyoid. The eyes have a tapetum lucidum. The inner margin of each pelvic fin in the male fish is grooved to constitute a clasper for the transmission of sperm. These fish are widely distributed in tropical and temperate waters.

Many fish maintain buoyancy with swim bladders. However elasmobranchs lack swim bladders, and maintain buoyancy instead with large livers that are full of oil. This stored oil may also function as a nutrient when food is scarce. Deep sea sharks are usually targeted for their oil, because the livers of these species can weigh up to 20% of their total weight.

Evolution

Fossilised shark teeth are known from the early Devonian, around 400 million years ago. During the following Carboniferous period, the sharks underwent a period of diversification, with many new forms evolving. Many of these became extinct during the Permian, but the remaining sharks underwent a second burst of adaptive radiation during the Jurassic, around which time the skates and rays first appeared. Many surviving orders of elasmobranch date back to the Cretaceous, or earlier.

Habitats

Elasmobranchs are mostly a marine taxon, but we know several species that live in freshwater environment (approximately 60 species which represent only 5% of the 1154 described species). They can be divided into two groups.

The euryhaline elasmobranchs, which are marine species that may survive and reproduce in freshwater environment, and the obligated freshwater elasmobranchs. The second group contains elasmobranchs that only lives in freshwater environment their entire life. This group contains only one clade: the subfamily Potamotrygoninae. This clade is endemic to one specific region (which means that they can only be seen in those regions): tropical, subtropical water and wetland of South America.

Recent researches in Paraná river have shown that obligated freshwater elasmobranchs were more susceptible to anthropogenic threats as overfishing and destruction of habitats due to the very small place they live in compared to the marin's species.

Holocephali

The subclass Holocephali ("complete heads") is a taxon of cartilaginous fish in the class Chondrichthyes. The earliest fossils are of teeth and come from the Devonian period. Little is known about these primitive forms, and the only surviving group in the subclass is the order Chimaeriformes. This group includes the rat fishes in the genus Chimaera, and the elephant fishes in the genus Callorhynchus. These fishes move by using sweeping movements of their large pectoral fins. They have long slender tails and live close to the seabed feeding on benthic invertebrates. They lack a stomach, food moving directly into the intestine.

Characteristics

Members of this taxon preserve today some features of elasmobranch life in Paleozoic times, though in other respects they are aberrant. They live close to the bottom and feed on molluscs and other invertebrates. The tail is long and thin and they move by sweeping movements of the large pectoral fins. The erectile spine in front of the dorsal fin is sometimes poisonous. There is no stomach (that is, the gut is simplified and the 'stomach' is merged with the intestine), and the mouth is a small aperture surrounded by lips, giving the head a parrot-like appearance. The only surviving members of the group are the rabbit fish (Chimaera), and the elephant fishes (Callorhynchus).

Evolution

The fossil record of the Holocephali starts during the Devonian period. The record is extensive, but most fossils are teeth, and the body forms of numerous species are not known, or at best poorly understood. Some experts[who?] further group the orders Petalodontiformes, Iniopterygiformes, and Eugeneodontida into the taxon "Paraselachimorpha", and treat it as a sister group to Chimaeriformes. However, as almost all members of Paraselachimorpha are poorly understood, most experts suspect this taxon to be either paraphyletic or a wastebasket taxon.

Batoidea

Batoidea is a superorder of cartilaginous fish commonly known as rays. They and their close relatives, the sharks, comprise the subclass Elasmobranchii. Rays are the largest group of cartilaginous fishes, with well over 600 species in 26 families. Rays are distinguished by their flattened bodies, enlarged pectoral fins that are fused to the head, and gill slits that are placed on their ventral surfaces.

Anatomy

Batoids are flat-bodied, and, like sharks, are cartilaginous marine fish, meaning they have a boneless skeleton made of a tough, elastic substance. Most batoids have five ventral slot-like body openings called gill slits that lead from the gills, but the Hexatrygonidae have six. Batoid gill slits lie under the pectoral fins on the underside, whereas a shark's are on the sides of the head. Most batoids have a flat, disk-like body, with the exception of the guitarfishes and sawfishes, while most sharks have a spindle-shaped body. Many species of batoid have developed their pectoral fins into broad flat wing-like appendages. The anal fin is absent. The eyes and spiracles are located on top of the head. Batoids have a ventrally located mouth and can considerably protrude their upper jaw (palatoquadrate cartilage) away from the cranium to capture prey. The jaws have euhyostylic type suspension, which relies completely on the hyomandibular cartilages for support. Bottom-dwelling batoids breathe by taking water in through the spiracles, rather than through the mouth as most fishes do, and passing it outward through the gills.

Reproduction

Batoids reproduce in a number of ways. As is characteristic of elasmobranchs, batoids undergo internal fertilisation. Internal fertilisation is advantageous to batoids as it conserves sperm, does not expose eggs to consumption by predators, and ensures that all the energy involved in reproduction is retained and not lost to the environment. All skates and some rays are oviparous (egg laying) while other rays are ovoviviparous, meaning that they give birth to young which develop in a womb but without involvement of a placenta.

The eggs of oviparous skates are laid in leathery egg cases that are commonly known as mermaid's purses and which often wash up empty on beaches in areas where skates are common.

Habitat

Most species live on the sea floor, in a variety of geographical regions — mainly in coastal waters, although some live in deep waters to at least 3,000 metres (9,800 ft). Most batoids have a cosmopolitan distribution, preferring tropical and subtropical marine environments, although there are temperate and cold-water species. Only a few species, like manta rays, live in the open sea, and only a few live in freshwater, while some batoids can live in brackish bays and estuaries.

Feeding

Most batoids have developed heavy, rounded teeth for crushing the shells of bottom-dwelling species such as snails, clams, oysters, crustaceans, and some fish, depending on the species. Manta rays feed on plankton.

Classification

Extant orders of cartilaginous fishes									
Order	Image	Common name	Families	Genera	Species				Comment
					Total	CR	EN	VU	
Mylio-bati-formes		Sting-rays and relatives	10	29	223	1	16	33	Myliobatiformes include stingrays, butterfly rays, eagle rays, and manta rays. They were formerly included in the order Rajiformes, but more-recent phylogenetic studies have shown that they are a monophyletic group, and that its more-derived members evolved their highly flattened shapes independently of the skates.
Rajif-ormes		Skates and relatives	5	36	270	4	12	26	Rajiformes include skates, guitarfishes, and wedgefishes. They are distinguished by the presence of greatly enlarged pectoral fins, which reach as far forward as the sides of the head, with a generally flattened body. The undulatory pectoral fin motion diagnostic to this taxon is known as rajiform locomotion. The eyes and spiracles are located on the upper surface of the body, and the gill slits on the underside. They have flattened, crushing teeth, and are generally carnivorous. Most species give birth to live young, although some lay eggs inside a protective capsule or mermaid's purse.
Torpe-dini-formes		Electric rays	4	12	69	2		9	The electric rays have electric organs in their pectoral fin discs that generate electric current. They are used to immobilize prey and for defense. The current is strong enough to stun humans, and the ancient Greeks and Romans used these fish to treat ailments such as headaches.
Rhino-pristi-formes		Shov-elnose rays and relatives	1	2	5-7	3-5	2		The sawfishes are shark-like in form, having tails used for swimming and smaller pectoral fins than most batoids. The pectoral fins are attached above the gills as in all batoids, giving the fishes a broad-headed appearance. They have long, flat snouts with a row of tooth-like projections on either side. The snouts are up to 1.8 metres (6 ft) long, and 30 centimetres (1 ft) wide, and are used for slashing and impaling small fishes and to probe in the mud for embedded animals. Sawfishes can enter freshwater rivers and lakes. Some species reach a total length of 6 metres (20 ft).

The classification of batoids is currently undergoing revision; however, molecular evidence refutes the hypothesis that skates and rays are derived sharks. Nelson's 2006 Fishes of the World recognizes four orders. The Mesozoic Sclerorhynchoidea are basal or incertae sedis; they show features of the Rajiformes but have snouts resembling those of sawfishes. However, evidence indicates they are probably the sister group to sawfishes.

Order Torpediniformes

- Family Hypnidae (coffin rays)

- Family Narcinidae (numbfishes)

- Family Narkidae (sleeper rays)

- Family Torpedinidae (torpedo rays)

Order Rhinopristiformes

- Family Glaucostegidae (giant guitarfishes)

- Family Platyrhinidae* (fanrays)

- Family Pristidae (sawfishes)

- Family Rhinidae (wedgefishes)

- Family Rhinobatidae (guitarfishes)

- Family Trygonorrhinidae (banjo rays)

- Family Zanobatidae* (panrays)

* the placement of these families is uncertain

Order Rajiformes

- Family Anacanthobatidae (legskates)

- Family Arhynchobatidae (softnose skates)

- Family Gurgesiellidae (pygmy skates)

- Family Rajidae (skates)

Order Myliobatiformes

- Family Aetobatidae (pelagic eagle rays)

- Family Dasyatidae (whiptail stingrays)

- Family Gymnuridae (butterfly rays)

- Family Hexatrygonidae (sixgill stingrays)

- Family Myliobatidae (devilrays)

- Family Plesiobatidae (giant stingarees)

- Family Potamotrygonidae (Neotropical stingrays)

- Family Rhinopteridae (cownose rays)

- Family Urolophidae (stingarees)

- Family Urotrygonidae (round stingrays)

Difference between Sharks and Rays

Sharks and rays are both cartilaginous fishes which can be contrasted with bony fishes. Rays are basically flattened sharks, adapted for feeding on the bottom. Guitarfish are somewhat between sharks and rays, and show characteristics of both (though they are classified as rays).

Comparison of sharks, guitar fishes and rays			
Characteristic	Shark	Guitar fish	Ray
Shape	laterally compressed spindle	dorsoventrally compressed (flattened) disc	dorsoventrally compressed (flattened) disc
Spiracles	not always present		always present
Habitat	usually pelagic surface feeders, though carpet sharks are demersal bottom feeders	demersal/pelagic mix	usually demersal bottom feeders
Eyes	usually at the side of the head	usually on top of the head	usually on top of the head
Gill openings	on the sides		ventral (underneath)
Pectoral fins	distinct	not distinct	not distinct
Tail	large caudal fin used for propulsion	caudal fin that can be used for propulsion	varies from thick tail as extension of body to a whip that can sting to almost no tail.
Locomotion	swim by moving their caudal (tail) fin from side to side	Guitar fish and sawfish have a caudal fin like the shark.	swim by flapping their pectoral fins like wings.

Agnatha

Agnatha is a superclass of jawless fish in the phylum Chordata, subphylum Vertebrata, consisting of both present (cyclostomes) and extinct (conodonts and ostracoderms) species. The group excludes all vertebrates with jaws, known as gnathostomes.

The agnathans as a whole are paraphyletic, because most extinct agnathans belong to the stem group of gnathostomes. Recent molecular data, both from rRNA and from mtDNA as well as embryological data strongly supports the hypothesis that living agnathans, the cyclostomes, are monophyletic.

The oldest fossil agnathans appeared in the Cambrian, and two groups still survive today: the lampreys and the hagfish, comprising about 120 species in total. Hagfish are considered members of the subphylum Vertebrata, because they secondarily lost vertebrae; before this event was inferred from molecular and developmental data, the group Craniata was created by Linnaeus (and is still sometimes used as a strictly morphological descriptor) to reference hagfish plus vertebrates. In addition to the absence of jaws, modern agnathans are characterised by absence of paired fins; the presence of a notochord both in larvae and adults; and seven or more paired gill pouches. Lampreys have a light sensitive pineal eye (homologous to the pineal gland in mammals). All living and most extinct Agnatha do not have an identifiable stomach or any appendages. Fertilization and development are both external. There is no parental care in the Agnatha class. The Agnatha are ectothermic or cold blooded, with a cartilaginous skeleton, and the heart contains 2 chambers.

While a few scientists still regard the living agnathans as only superficially similar, and argue that many of these similarities are probably shared basal characteristics of ancient vertebrates, recent classifications clearly place hagfish (the Myxini or Hyperotreti), with the lampreys (Hyperoartii) as being more closely related to each other than either is to the jawed fishes.

Metabolism

Agnathans are ectothermic, meaning they do not regulate their own body temperature. Agnathan metabolism is slow in cold water, and therefore they do not have to eat very much. They have no distinct stomach, but rather a long gut, more or less homogenous throughout its length. Lampreys feed on other fish and mammals. They rely on a row of sharp teeth to shred their host. Anticoagulant fluids preventing blood clotting are injected into the host, causing the host to yield more blood. Hagfish are scavengers, eating mostly dead animals. They also use a sharp set of teeth to break down the animal. The fact that Agnathan teeth are unable to move up and down limits their possible food types.

Body Covering

In modern agnathans, the body is covered in skin, with neither dermal or epidermal scales. The skin of hagfish has copious slime glands, the slime constituting their defense mechanism. The slime can sometimes clog up enemy fishes' gills, causing them to die. In direct contrast, many extinct agnathans sported extensive exoskeletons composed of either massive, heavy dermal armour or small mineralized scales.

Appendages

Almost all agnathans, including all extant agnathans, have no paired appendages, although most do have a dorsal or a caudal fin. Some fossil agnathans, such as osteostracans and pituriaspids, did have paired fins, a trait inherited in their jawed descendants.

Reproduction

Fertilization in lampreys is external. Mode of fertilization in hagfishes is not known. Development in both groups probably is external. There is no known parental care. Not much is known about the hagfish reproductive process. It is believed that hagfish only have 30 eggs over a lifetime. Most species are hermaphrodites. There is very little of the larval stage that characterizes the lamprey. Lamprey are only able to reproduce once. After external fertilization, the lamprey's cloacas remain open, allowing a fungus to enter their intestines, killing them. Lampreys reproduce in freshwater riverbeds, working in pairs to build a nest and burying their eggs about an inch beneath the sediment. The resulting hatchlings go through four years of larval development before becoming adults. They also have a certain unusual form of reproduction.

Evolution

Although a minor element of modern marine fauna, agnathans were prominent among the early fish in the early Paleozoic. Two types of Early Cambrian animal apparently having fins, vertebrate musculature, and gills are known from the early Cambrian Maotianshan shales of China: *Haikouichthys* and *Myllokunmingia*. They have been tentatively assigned to Agnatha by Janvier. A third possible agnathid from the same region is *Haikouella*. A possible agnathid that has not been formally described was reported by Simonetti from the Middle Cambrian Burgess Shale of British Columbia.

Many Ordovician, Silurian, and Devonian agnathans were armored with heavy bony-spiky plates. The first armored agnathans—the Ostracoderms, precursors to the bony fish and hence to the tetrapods (including humans)—are known from the middle Ordovician, and by the Late Silurian the agnathans had reached the high point of their evolution. Most of the ostracoderms, such as thelodonts, osteostracans, and galeaspids, were more closely related to the gnathostomes than to the surviving agnathans, known as cyclostomes. Cyclostomes apparently split from other agnathans before the evolution of dentine and bone, which are present in many fossil agnathans, including conodonts. Agnathans declined in the Devonian and never recovered.

Classification

Subgroups of jawless fish			
	Subgroup	Example	Comments
Cyclostomes	Myxini	hagfish	Myxini (hagfish) are eel-shaped slime-producing marine animals (occasionally called slime eels). They are the only known living animals that have a skull but not a vertebral column. Along with lampreys, hagfish are jawless and are living fossils; hagfish are basal to vertebrates, and living hagfish remain similar to hagfish 300 million years ago. The classification of hagfish has been controversial. The issue is whether the hagfish is itself a degenerate type of vertebrate-fish (most closely related to lampreys), or else may represent a stage which precedes the evolution of the vertebral column (as do lancelets). The original scheme groups hagfish and lampreys together as cyclostomes (or historically, Agnatha), as the oldest surviving clade of vertebrates alongside gnathostomes (the now-ubiquitous jawed-vertebrates). An alternative scheme proposed that jawed-vertebrates are more closely related to lampreys than to hagfish (i.e., that vertebrates include lampreys but exclude hagfish), and introduces the category craniata to group vertebrates near hagfish. Recent DNA evidence has supported the original scheme.

	Hyperoar-tia	lamprey	Hyperoartia is a disputed group of vertebrates that includes the modern lampreys and their fossil relatives. Examples of hyperoartians from early in their fossil record are *Endeiolepis* and *Euphanerops*, fish-like animals with hypocercal tails that lived during the Late Devonian Period. Some paleontologists still place these forms among the "ostracoderms" (jawless armored "fishes") of the class Anaspida, but this is increasingly considered an artificial arrangement based on ancestral traits. Placement of this group among the jawless vertebrates is a matter of dispute. While today enough fossil diversity is known to make a close relationship among the "ostracoderms" unlikely, this has muddied the issue of the Hyperoartia's closest relatives. Traditionally the group was placed in a superclass Cyclostomata together with the Myxini (hagfishes). More recently, it has been proposed that the Myxini are more basal among the skull-bearing chordates, while the Hyperoartia are retained among vertebrates. But even though this may be correct, the lampreys represent one of the oldest divergences of the vertebrate lineage, and whether they are better united with some "ostracoderms" in the Cephalaspidomorphi, or not closer to these than to e.g. to other "ostracoderms" of the Pteraspidomorphi, or even the long-extinct conodonts, is still to be resolved. Even the very existence of the Hyperoartia is disputed, with some analyses favoring a treatment of the "basal Hyperoartia" as a monophyletic lineage Jamoytiiformes that may in fact be very close to the ancestral jawed vertebrates.
Ostrac-oderms	**†Pteraspi-domorphi (extinct)**		†Pteraspidomorphi is an extinct group of early jawless fish. The fossils show extensive shielding of the head. Many had hypocercal tails in order to generate lift to increase ease of movement through the water for their armoured bodies, which were covered in dermal bone. They also had sucking mouth parts and some species may have lived in fresh water. The taxon contains the subgroups Heterostraci, Astraspida, Arandaspida.
	†Thelodonti (extinct)		Thelodonti *(nipple teeth)* are a group of small, extinct jawless fishes with distinctive scales instead of large plates of armour. There is much debate over whether the group of Palaeozoic fish known as the Thelodonti (formerly coelolepids) represent a monophyletic grouping, or disparate stem groups to the major lines of jawless and jawed fish. Thelodonts are united in possession of "thelodont scales". This defining character is not necessarily a result of shared ancestry, as it may have been evolved independently by different groups. Thus the thelodonts are generally thought to represent a polyphyletic group, although there is no firm agreement on this point; if they are monophyletic, there is no firm evidence on what their ancestral state was. "Thelodonts" were morphologically very similar, and probably closely related, to fish of the classes Heterostraci and Anaspida, differing mainly in their covering of distinctive, small, spiny scales. These scales were easily dispersed after death; their small size and resilience makes them the most common vertebrate fossil of their time. The fish lived in both freshwater and marine environments, first appearing during the Ordovician, and perishing during the Frasnian–Famennian extinction event of the Late Devonian. They were predominantly deposit-feeding bottom dwellers, although there is evidence to suggest that some species took to the water column to be free-swimming organisms.

	†**Anaspida (extinct)**		Anaspida *(without shield)* is an extinct group of primitive jawless vertebrates that lived during the Silurian and Devonian periods. They are classically regarded as the ancestors of lampreys. Anaspids were small marine agnathans that lacked heavy bony shield and paired fins, but have a striking highly hypocercal tail. They first appeared in the early Silurian, and flourished until the Late Devonian extinction, where most species, save for lampreys, became extinct due to the environmental upheaval during that time.
	†**Cephala-spido-morphi (extinct)**		Cephalaspidomorphi is a broad group of extinct armored agnathans found in Silurian and Devonian strata of North America, Europe, and China, and is named in reference to the osteostracan genus *Cephalaspis*. Most biologists regard this taxon as extinct, but the name is sometimes used in the classification of lampreys, as lampreys are sometimes thought to be related to cephalaspids. If lampreys are included, they would extend the known range of the group from the early Silurian period through the Mesozoic, and into the present day. Cephalaspidomorphi were, like most contemporary fish, very well armoured. Particularly the head shield was well developed, protecting the head, gills and the anterior section of the innards. The body was in most forms well armoured as well. The head shield had a series of grooves over the whole surface forming an extensive lateral line organ. The eyes were rather small and placed on the top of the head. There was no proper jaw. The mouth opening was surrounded by small plates making the lips flexible, but without any ability to bite. Undisputed subgroups traditionally contained with Cephaloaspidomorphi, also called "Monorhina," include the classes Osteostraci, Galeaspida, and Pituriaspida

Groups

- Cyclostomes
    - o Myxini (hagfish)
    - o Hyperoartia (Petromyzontida)
        - ▪ Petromyzontidae (lampreys)
- Ostracoderms
    - o †Pteraspidomorphi
    - o †Thelodonti
    - o †Anaspida
    - o Cephalaspidomorphi
        - ▪ †Galeaspida
        - ▪ †Pituriaspida
        - ▪ †Osteostraci

Cyclostomata

Cyclostomata is a group of agnathans that comprises the living jawless fishes: the lampreys and hagfishes. Both groups have round mouths that lack jaws but have retractable horny teeth. The name Cyclostomata means "round mouths". Their mouths cannot close due to the lack of a jaw, so they have to constantly cycle water through the mouth.

Possible Relationships

This taxon is often included in the paraphyletic superclass Agnatha, which also includes several groups of extinct armored fishes called ostracoderms. Most fossil agnathans, such as galeaspids, thelodonts, and osteostracans, are more closely related to vertebrates with jaws (called gnathostomes) than to cyclostomes. Cyclostomes seem to have split off before the evolution of dentine and bone, which are present in many fossil agnathans, including conodonts.

Biologists disagree about whether cyclostomes are a clade. The "vertebrate hypothesis" holds that lampreys are more closely related to gnathostomes than they are to the hagfish. The "cyclostome hypothesis", on the other hand, holds that lampreys and hagfishes are more closely related, making cyclostomata monophyletic.

Most studies based on anatomy have supported the vertebrate hypothesis, while most molecular phylogenies have supported the cyclostome hypothesis.

There are exceptions in both cases, however. Similarities in the cartilage and muscles of the tongue apparatus also provide evidence of sister-group relationship between lampreys and hagfishes. And at least one molecular phylogeny has supported the vertebrate hypothesis. The embryonic development of hagfishes was once held to be drastically different from that of lampreys and gnathostomes, but recent evidence suggests that it is more similar than previously thought, which may remove an obstacle to the cyclostome hypothesis. There is at present no consensus on the correct topology.

Differences and Similarities

Both hagfishes and lampreys have just one gonad, but for different reasons. In hagfishes it is because only a single gonad is developed during their ontogeny, while it is achieved through the fusion of gonads in lampreys.

Unlike jawed vertebrates, which have three semicircular canals in each inner ear, lampreys have only two and hagfishes just one. But the semicircular canal of hagfishes contains both stereocilia and a second class of hair cells, apparently a derived trait, whereas lampreys and other vertebrates have stereocilia only. Because the inner ear of hagfishes has two forms of sensory ampullae, their single semicircular canal is assumed to be a result of two semicircular canals that have merged into just one.

The hagfish blood is isotonic with seawater, while lampreys appears to use the same gill-based mechanisms of osmoregulation as marine teleosts. Yet the same mechanisms are apparent in the mitochondria-rich cells in the gill epithelia of hagfishes, but never develops the ability to regulate the blood's salinity, even if they are capable of regulating the ionic concentration of Ca and Mg ions.

The lamprey intestine has a typhlosole that increases the inner surface like the spiral valve does in some jawed vertebrates. The spiral valve in the latter develops by twisting the whole gut, while the lamprey typhlosole is confined to the mucous membrane of the intestines. The mucous membranes of hagfishes have a primitive typhlosole in the form of permanent zigzag ridges. This trait could be a primitive one, since it is also found in some sea squirts such as *Ciona*. The intestinal epiphelia of lampreys also have ciliated cells, which have not been detected in hagfishes. Because ciliated intestines are also found in Chondrostei, lungfishes and the early stages of some teleosts, it is considered a primitive condition that has been lost in hagfishes.

Hagfish

Hagfish, the class Myxini (also known as Hyperotreti), are eel-shaped, slime-producing marine fish (occasionally called slime eels). They are the only known living animals that have a skull but no vertebral column, although hagfish do have rudimentary vertebrae. Along with lampreys, hagfish are jawless; they are the sister group to vertebrates, and living hagfish remain similar to hagfish from around 300 million years ago.

The classification of hagfish has been controversial. The issue is whether the hagfish is a degenerate type of vertebrate-fish (most closely related to lampreys), or represents a stage that precedes the evolution of the vertebral column (as do lancelets). The original scheme groups hagfish and lampreys together as cyclostomes (or historically, Agnatha), as the oldest surviving class of vertebrates alongside gnathostomes (the now-ubiquitous jawed vertebrates). An alternative scheme proposed that jawed vertebrates are more closely related to lampreys than to hagfish (i.e., that vertebrates include lampreys but exclude hagfish), and introduces the category craniata to group vertebrates near hagfish. Recent DNA evidence has supported the original scheme.

Physical Characteristics

Body Features

Hagfish average about 0.5 m (19.7 in) in length. The largest known species is *Eptatretus goliath* with a specimen recorded at 127 cm (4 ft 2 in), while *Myxine kuoi* and *Myxine pequenoi* seem to reach no more than 18 cm (7.1 in) (some have been seen as small as 4 cm (1.6 in)).

Pacific hagfish at 150 m depth, California, Cordell Bank National Marine Sanctuary

Hagfish have elongated, eel-like bodies, and paddle-like tails. The skin is naked and covers the body like a loosely fitting sock. They have cartilaginous skulls (although the part surrounding the brain is composed primarily of a fibrous sheath) and tooth-like structures composed of keratin. Colors depend on the species, ranging from pink to blue-grey, and black or white spots may be present. Eyes are simple eyespots, not compound eyes that can resolve images. Hagfish have no true fins and have six or eight barbels around the mouth and a single nostril. Instead of vertically articulating jaws like Gnathostomata (vertebrates with jaws), they have a pair of horizontally moving structures with tooth-like projections for pulling off food. The mouth of the hagfish has two pairs of horny, comb-shaped teeth on a cartilaginous plate that protracts and retracts. These teeth are used to grasp food and draw it toward the pharynx.

Slime

Pacific hagfish trying to hide under a rock

Hagfish are long and vermiform, and can exude copious quantities of a milky and fibrous slime or mucus from some 100 glands or invaginations running along their flanks. The typical species *Myxine glutinosa* was named for this slime. When captured and held, e.g., by the tail, they secrete the microfibrous slime, which expands into up to 20 litres (5¼ gallons) of sticky, gelatinous material when combined with water. If they remain captured, they can tie themselves in an overhand knot, which works its way from the head to the tail of the animal, scraping off the slime as it goes and freeing them from their captor. This singular behavior may assist them in extricating themselves from the jaws of predatory fish or from the interior of their own "prey", and the "sliming" might act as a distraction to predators.

Recently, the slime was reported to entrain water in its microfilaments, creating a slow-to-dissipate, viscoelastic substance, rather than a simple gel. It has been proven to impair the function of a predator fish's gills. In this case, the hagfish's mucus would clog up the predator's gills, disabling their ability to breathe. The predator would release the hagfish to avoid suffocation. It is because of the mucus that there are very few marine predators that target the hagfish. The other known predators of hagfish are varieties of birds or mammals.

Free-swimming hagfish also "slime" when agitated and later clear the mucus off by way of the same travelling-knot behavior. The reported gill-clogging effect suggests that the travelling-knot behavior is useful or even necessary to restore the hagfish's own gill function after "sliming".

Research is ongoing regarding the properties and possible applications of the components of hagfish slime filament protein, particularly as a renewable alternative to synthetics currently derived from petroleum.

Respiration

Hagfish generally respire through taking in water through their pharynx, past the velar chamber and bringing the water through the internal gill pouches, which can vary in number from 5 to 16 pairs, depending on species. The gill pouches open individually, but in Myxine the openings have coalesced, with canals running backwards from each opening under the skin, uniting to form a common aperture on the ventral side known as the branchial opening. The esophagus is also connected to the left branchial opening, which is therefore larger than the right one, through a pharyngocutaneous duct (esophageocutaneous duct), which has no respiratory tissue. This pharyngocutaneous duct is used to clear large particles from the pharynx, a function also partly taking place through the nasopharyngeal canal. In other species the coalescence of the gill openings is less complete, and in Bdellostoma each pouch opens separately to the outside like in lampreys. The unidirectional water flow passing the gills is produced by rolling and unrolling velar folds located inside a chamber developed from the naso-hypophyseal tract, and is operated by a complex set of muscles inserting into cartilages of the neurocranium, assisted by peristaltic contractions of the gill pouches and their ducts. Hagfish also have a well-developed dermal capillary network that supplies the skin with oxygen when the animal is buried in anoxic mud, as well as a high tolerance for both hypoxia and anoxia, with a well developed anaerobic metabolism. It has also been suggested that the skin is capable of cutaneous respiration.

Eye

The hagfish's eye, which lacks lens, extraocular muscles, and the three motor cranial nerves (III, IV, and VI), is significant to the evolution of more complex eyes. A parietal eye and the parapineal organ are also absent. Hagfish eyespots, when present, can detect light, but as far as is known, none can resolve detailed images. In *Myxine* and *Neomyxine*, the eyes are partly covered by the trunk musculature.

Reproduction

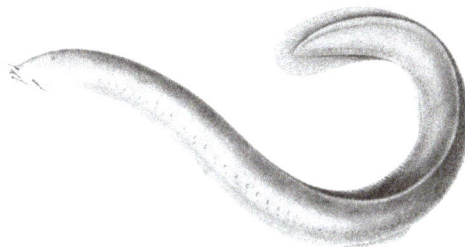

Drawing of *Eptatretus polytrema*

Very little is known about hagfish reproduction. Embryos are difficult to obtain for study, although laboratory breeding of the Far Eastern inshore hagfish, *Eptatretus burgeri*, has succeeded. In some species, sex ratio has been reported to be as high as 100:1 in favor of females. Some hagfish species are thought to be hermaphroditic, having both an ovary and a testicle (there is only one gamete production organ in both females and males). In some cases, the ovary is thought to remain non-functional until the individual has reached a particular age or encounters a particular environmental stress. These two factors in combination suggest the survival rate of hagfish is quite high.

Drawing of a New Zealand hagfish

Depending on species, females lay from one to thirty tough, yolky eggs. These tend to aggregate due to having Velcro-like tufts at either end. Hagfish are sometimes seen curled around small clutches of eggs. It is not certain if this constitutes actual breeding behavior.

Hagfish do not have a larval stage, in contrast to lampreys, which have a long one.

Hagfish have a mesonephric kidney and are often neotenic of their pronephric kidney. The kidney(s) are drained via mesonephric/archinephric duct. Unlike many other vertebrates, this duct is separate from the reproductive tract. Unlike all other vertebrates, the proximal tubule of the nephron is also connected with the coelom, provided lubrication.

The single testicle or ovary has no transportation duct. Instead, the gametes are released into the coelom until they find their way to the posterior end of the caudal region, whereby they find an opening in the digestive system.

Feeding

While polychaete marine worms on or near the sea floor are a major food source, hagfish can feed upon and often even enter and eviscerate the bodies of dead and dying/injured sea creatures much larger than themselves. They are known to devour their prey from the inside. Hagfish have the ability to absorb dissolved organic matter across the skin and gill, which may be an adaptation to a scavenging lifestyle, allowing them to maximize sporadic opportunities for feeding. From an evolutionary perspective, hagfish represent a transitory state between the generalized nutrient absorption pathways of aquatic invertebrates and the more specialized digestive systems of aquatic vertebrates.

Two Pacific hagfish feeding on a dead sharpchin rockfish, *Sebastes zacentrus*, while one remains in a curled position at the left of the photo

Like leeches, they have a sluggish metabolism and can survive months between feedings; their feeding behavior, however, appears quite vigorous. Analysis of the stomach content of several species has revealed a large variety of prey, including polychaetes, shrimps, hermit crabs, cephalopods, brittlestars, bony fishes, sharks, birds and whale flesh.

In captivity, hagfish are observed to use the overhand-knot behavior "in reverse" (tail-to-head) to assist them in gaining mechanical advantage to pull out chunks of flesh from carrion fish or cetaceans, eventually making an opening to permit entry to the interior of the body cavity of larger carcasses. A healthy larger sea creature likely would be able to outfight or outswim this sort of assault.

This energetic opportunism on the part of the hagfish can be a great nuisance to fishermen, as they can devour or spoil entire deep-drag-netted catches before they can be pulled to the surface. Since hagfish are typically found in large clusters on and near the bottom, a single trawler's catch could contain several dozen or even hundreds of hagfish as bycatch, and all the other struggling, captive sea life make easy prey for them.

The digestive tract of the hagfish is unique among the chordates because the food in the gut is enclosed in a permeable membrane, analogous to the peritrophic matrix of insects.

Hagfish have also been observed actively hunting the red bandfish, *Cepola haastii*, in its burrow, possibly using their slime to suffocate the fish before grasping it with their dental plates and dragging it from the burrow.

Classification

Originally, *Myxine* was included by Linnaeus (1758) in Vermes. In recent years, hagfish have become of special interest for genetic analysis investigating the relationships among chordates. Their classification as agnathans places hagfish as elementary vertebrates in between invertebrates and gnathostomes. However, there has been long discussion in scientific literature about whether the hagfish were even non-vertebrate. This position is supported by recent molecular biology analyses, which tend to classify hagfish as invertebrates within the subphylum Craniata, because of their molecular evolutionary distance from Vertebrata (*sensu stricto*). A single fossil of hagfish shows little evolutionary change has occurred in the last 300 million years.

However, the validity of the taxon "Craniata" was recently examined by Delarbre et al. (2002) using mtDNA sequence data, concluding the Myxini are more closely related to Hyperoartia than to Gnathostomata – i.e., that modern jawless fishes form a clade called Cyclostomata. The argument is that if the Cyclostomata are indeed monophyletic, Vertebrata would return to its old content (Gnathostomata + Cyclostomata) and the name Craniata, being superfluous, would become a junior synonym. The current classification supported by molecular analyses (which show that lampreys and hagfishes are sister taxa), as well as the fact that hagfishes do, in fact, have rudimentary vertebrae places hagfishes in Cyclostomata.

Commercial Use

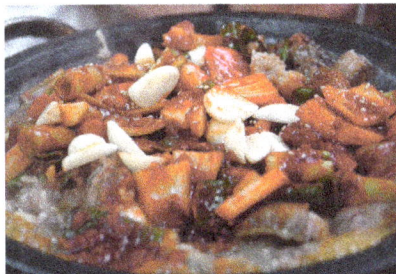

Kkomjangeo bokkeum, Korean stir-fried fish dish made with the hagfish *Eptatretus burgeri*

As Human Food

Hagfish are not often eaten, owing to their repugnant looks and sliminess. However the inshore hagfish, found in the Northwest Pacific, is valued as food in Korea. The hagfish is kept alive and irritated by rattling its container with a stick, prompting it to produce slime in large quantities. This slime is used in a similar manner as egg whites in various forms of cookery in the region. The inshore hagfish, known as *kkomjangeo* or *meokjango* in Korean and *nuta-unagi* in Japanese, is the only member of the hagfish family with a seasonal reproductive cycle.

Skins

Hagfish skin, used in a variety of clothing accessories, is usually referred to as eel skin. It produces a particularly durable leather, especially suitable for wallets and belts.

Hyperoartia

Hyperoartia or Petromyzontida is a disputed group of vertebrates that includes the modern lampreys and their fossil relatives. Examples of hyperoartians from early in their fossil record are *Endeiolepis* and *Euphanerops*, fish-like animals with hypocercal tails that lived during the Late Devonian Period. Some paleontologists still place these forms among the "ostracoderms" (jawless armored fishes) of the class Anaspida, but this is increasingly considered an artificial arrangement based on ancestral traits.

Placement of this group among the jawless vertebrates is a matter of dispute. While today enough fossil diversity is known to make a close relationship among the "ostracoderms" unlikely, this has muddied the issue of the Hyperoartia's closest relatives. Traditionally the group was placed in a superclass Cyclostomata together with the Myxini (hagfishes). More recently, it has been proposed that the Myxini are more basal among the skull-bearing chordates, while the Hyperoartia are retained among vertebrates. But even though this may be correct, the lampreys represent one of the oldest divergences of the vertebrate lineage, and whether they are better united with some "ostracoderms" in the Cephalaspidomorphi, or not closer to these than to e.g. to other "ostracoderms" of the Pteraspidomorphi, or even the long-extinct conodonts, is still to be resolved. Even the very existence of the class Hyperoartia is disputed, with some analyses favoring a treatment of the "basal Hyperoartia" as a monophyletic lineage Jamoytiiformes that may in fact be very close to the ancestral jawed vertebrates.

The only hyperoartians surviving today are lampreys, classified in the Petromyzontiformes. The discovery of the fossil *Priscomyzon* pushed back the oldest known occurrence of true lampreys to the Late Devonian. The evidence of phylogeny, however, suggests the lamprey lineage diverged much earlier from other vertebrates, rather than arising from among the "ostracoderms". The origin of Hyperoartia may therefore extend back to the early Paleozoic, if not earlier.

Mayomyzon pickoensis

Placed in this group are at present:

- †Jamoytiiformes

    o   †Jamoytiidae White 1946

        ▪   †*Jamoytius kerwoodi* White 1946

- †Endeiolepidiformes Berg 1940

    o   †Endeiolepididae Stensio 1939

        ▪   †*Endeiolepis aneri* Stensio 1939

- †Euphanerida

    o   †Euphaneropidae Woodward 1900

        ▪   †*Euphanerops longaevus* Woodward 1900 [*Legendrelepis* Arsenault & Janvier 1991; *Legendrelepis parenti* Arsenault & Janvier 1991]

        ▪   †*Cornovichthys blaauweni* Newman & Trewin 2001

- †*Haikouichthys ercaicunensis* Luo, Hu & Shu 1999; sensu Shu et al. 2003

- Petromyzontimorpha

    o   Petromyzontiformes (Lampreys)

Ostracoderm

Various ostracoderms of the class Osteostraci *(bony shields)*

Ostracoderms ("shell-skinned") are the armored jawless fishes of the Paleozoic. The term does not often appear in classifications today because it is paraphyletic or polyphyletic, and has no phylogenetic meaning. However, the term is still used as an informal way of loosely grouping together the armored jawless fishes.

An innovation of ostracoderms was the use of gills not for feeding, but exclusively for respiration. Earlier chordates with gills used them for both respiration and feeding. Ostracoderms had separate pharyngeal gill pouches along the side of the head, which were permanently open with no protective operculum. Unlike invertebrates that use ciliated motion to move food, ostracoderms used their muscular pharynx to create a suction that pulled small and slow moving prey into their mouths.

Cardipeltis bryanti, a lower Devonian ostracoderm from the Bighorn Mountains of Wyoming. Ventral (underside) exposed.

The first fossil fishes that were discovered were ostracoderms. The Swiss anatomist Louis Agassiz received some fossils of bony armored fish from Scotland in the 1830s. He had a hard time classifying them as they did not resemble any living creature. He compared them at first with extant armored fish such as catfish and sturgeons but later realizing that they had no movable jaws, classified them in 1844 into a new group "ostracoderms" which means "shell-skinned".

Ostracoderms have heads covered with a bony shield. They are among the earliest creatures with bony heads. The microscopic layers of that shield appear to evolutionary biologists, "like they are composed of little tooth-like structures." Neil Shubin writes: "Cut the bone of the [ostracoderm] skull open...pop it under a microscope and...you find virtually the same structure as in our teeth. There is a layer of enamel and even a layer of pulp. The whole shield is made up of thousands of small teeth fused together. This bony skull--one of the earliest in the fossil record--is made entirely of little teeth. Teeth originally arose to bite creatures (see Conodonts); later a version of teeth was used in a new way to protect them."

Ostracoderms existed in two major groups, the more primitive heterostracans and the cephalaspids. The cephalaspids were more advanced than the heterostracans in that they had lateral stabilizers for more control of their swimming.

It was long assumed that pteraspidomorphs and thelodonts were the only ostracoderms with paired nostrils, while the other groups have just a single median nostril. But it has since been revealed that even if galeaspidans have just one external opening, it has two internal nasal organs.

After the appearance of jawed fish (placoderms, acanthodians, sharks, etc.) about 420 million years ago, most ostracoderm species underwent a decline, and the last ostracoderms became extinct at the end of the Devonian period. More recent research indicates, however, that fish with jaws had far less to do with the extinction of the ostracoderms than previously assumed, as they coexisted without noticeable decline for about 30 million years.

The Subclass Ostracodermi has been placed in the division Agnatha along with the extant Subclass Cyclostomata, which includes lampreys and hagfishes.

Major groups

Major groups of ostracoderms				
Group	Class	Image	Description	
Cephalaspi-do-morphi			Cephalaspidomorphi, or cephalaspids, were like most contemporary fishes very well armoured. Particularly the head shield was well developed, protecting the head, gills and the anterior section of innards. The body were in most forms well armoured too. The head shield had a series of grooves over the whole surface forming an extensive lateral line organ. The eyes were rather small and placed on the top of the head. There was no jaw proper. The mouth opening was surrounded by small plates making the lips flexible, but without any ability to bite. Most biologists regard this taxon as extinct, but the name is sometimes used in the classification of lampreys because lampreys were once thought to be related to cephalaspids. If lampreys are included, they would extend the known range of the group from the Silurian and Devonian periods to the present day.	
	Galeaspida (extinct)		Galeaspida *(helmet shields)* have massive bone shield on the head. Galeaspida lived in shallow, fresh water and marine environments during the Silurian and Devonian times (430 to 370 million years ago) in what is now Southern China, Tibet and Vietnam. Superficially, their morphology appears more similar to that of Heterostraci than Osteostraci, there being currently no evidence that the galeaspids had paired fins. However, Galeaspida are in fact regarded as being more closely related to Osteostraci, based on the closer similarity of the morphology of the braincase.	
	Pituriaspida (extinct)		Pituriaspida *(pituri or hallucinogenic shield)* are a small group of extinct armoured jawless fishes with tremendous nose-like rostrums, which lived in the marine, deltaic environments of Middle Devonian Australia (about 390 Ma). They are known only by two species, *Pituriaspis doylei* and *Neeyambaspis enigmatica* found in a single sandstone location of the Georgina Basin, in Western Queensland, Australia	
	Osteostra-ci (extinct)		Osteostraci *(bony shields)* lived in what is now North America, Europe and Russia from the Middle Silurian to Late Devonian. Anatomically speaking, the osteostracans, especially the Devonian species, were among the most advanced of all known agnathans. This is due to the development of paired fins, and their complicated cranial anatomy. The osteostracans were more similar to lampreys than to jawed vertebrates in possessing two pairs of semicircular canals in the inner ear, as opposed to the three pairs found in the inner ears of jawed vertebrates. They are thought to be the sister-group of pituriaspids, and together, these two taxa of jawless vertebrates are the sister-group of gnathostomes. Several synapomorphies support this hypothesis, such as the presence of: sclerotic ossicles, paired pectoral fins, a dermal skeleton with three layers (a basal layer of isopedin, a middle layer of spongy bone, and a superficial layer of dentin), and perichondral bone.	

Other groups			
Other groups	†**Pteraspi-do-morphi (extinct)**		†Pteraspidomorphi have extensive shielding of the head. Many had hypocercal tails in order to generate lift to increase ease of movement through the water for their armoured bodies, which were covered in dermal bone. They also had sucking mouth parts and some species may have lived in fresh water. The taxon contains the subgroups Heterostraci, Astraspida, Arandaspida.
	†**Thelodonti (extinct)**		Thelodonti *(nipple teeth)* are a group of small, extinct jawless fishes with distinctive scales instead of large plates of armour. There is much debate over whether the group of Palaeozoic fish known as the Thelodonti (formerly coelolepids) represent a monophyletic grouping, or disparate stem groups to the major lines of jawless and jawed fish. Thelodonts are united in possession of "thelodont scales". This defining character is not necessarily a result of shared ancestry, as it may have been evolved independently by different groups. Thus the thelodonts are generally thought to represent a polyphyletic group, although there is no firm agreement on this point; if they are monophyletic, there is no firm evidence on what their ancestral state was. "Thelodonts" were morphologically very similar, and probably closely related, to fish of the classes Heterostraci and Anaspida, differing mainly in their covering of distinctive, small, spiny scales. These scales were easily dispersed after death; their small size and resilience makes them the most common vertebrate fossil of their time. The fish lived in both freshwater and marine environments, first appearing during the Ordovician, and perishing during the Frasnian–Famennian extinction event of the Late Devonian. They were predominantly deposit-feeding bottom dwellers, although there is evidence to suggest that some species took to the water column to be free-swimming organisms.
	†**Anaspida (extinct)**		Anaspida *(without shield)* is an extinct group of primitive jawless vertebrates that lived during the Silurian and Devonian periods. They are classically regarded as the ancestors of lampreys. Anaspids were small marine agnathans that lacked heavy bony shield and paired fins, but have a striking highly hypocercal tail. They first appeared in the early Silurian, and flourished until the Late Devonian extinction, where most species, save for lampreys, became extinct due to the environmental upheaval during that time.

Thelodonti

Thelodonti is a class of extinct jawless fishes with distinctive scales instead of large plates of armor.

There is much debate over whether the group of Palaeozoic fish known as the Thelodonti (formerly coelolepids) represent a monophyletic grouping, or disparate stem groups to the major lines of jawless and jawed fish.

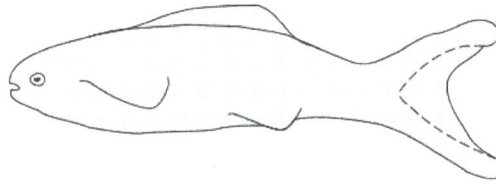

Shielia taiti

Thelodonts are united in possession of "thelodont scales". This defining character is not neces-sarily a result of shared ancestry, as it may have been evolved independently by different groups. Thus the thelodonts are generally thought to represent a polyphyletic group, although there is no firm agreement on this point; if they are monophyletic, there is no firm evidence regarding their ancestral state.

"Thelodonts" were morphologically very similar, and probably closely related, to fish of the classes Heterostraci and Anaspida, differing mainly in their covering of distinctive, small, spiny scales. These scales were easily dispersed after death; their small size and resilience makes them the most common vertebrate fossil of their time.

The fish lived in both freshwater and marine environments, first appearing during the Ordovician, and perishing during the Frasnian–Famennian extinction event of the Late Devonian. They were predominantly deposit-feeding bottom dwellers, although there is evidence to suggest that some species took to the water column to be free-swimming organisms.

Description

Very few complete thelodont specimens are known; fewer still are preserved in three dimensions. This is due in part to the lack of an internal ossified (i.e. bony) skeleton; it does not help that the scales are poorly, if at all, attached to one another, and that they readily detach from their owners upon death.

The exoskeleton is composed of many tooth-like scales, usually around 0.5-1.5mm in size. These scales did not overlap, and were aligned to point backwards along the fish, in the most streamlined direction, but beyond that, often appear haphazard in their orientation. The scales themselves ap-proximate the form of a teardrop mounted on a small, bulky base, with the base often containing a small rootlet with which the scale was attached to the fish. The "teardrop" often contains lines, ridges, furrows and spikes running down its length in an array of sometimes complex patterns. Scales found around the gill region were generally smaller than the larger, bulkier scales found on the dorsal/ventral sides of the fish; some genera display rows of longer spikes.

The scaly covering contrasts them with most other jawless fishes, which were armor-plated with large, flat scales.

Aside from scattered scales, some specimens do appear to display imprints, giving an indication of the structure of the whole animal - which appeared to reach 15–30 cm in length. Tentative studies appear to suggest that the fish possessed a more developed braincase than the lampreys, with an almost shark-like outline. Internal scales have also been recovered, some fused into plates resembling gnathostome tooth-whorls to such a degree that some researchers favour a close link between the families.

Despite the rarity of complete fossils, these very rare intact specimens do allow us to gain an insight into the internal organ arrangement of the Thelodonts. Some specimens described in 1993 were the first to be found with a significant degree of three-dimensionality, ending speculations that the Thelodonts were flattened fish. Further, these fossils allowed the gut morphology to be interpreted, which generated much excitement: their guts were unlike those of any other agnathans, and a stomach was clearly visible: this was unexpected, as it was previously thought that stomachs evolved *after* jaws. Distinctive fork-shaped tails - usually characteristic of the jawed fish (gnathostomes) - were also found, linking the two groups to an unexpected degree.

The fins of the thelodonts are useful in reconstructing their mode of life. Their paired pectoral fins were combined with a single, usually well-developed, dorsal and anal fins; these and the hypercercal and much larger hypocercal lobes forming a heterocercal tail resemble features of modern fish that are associated with their deftness at predation and evasion.

Taxonomy

Due to the small number of intact fossils, the taxonomy of thelodonts is based primarily on scale morphology. In fact, some thelodont families are only recognised based on their scale fossils.

A recent assessment of thelodont taxonomy by Wilson and Märss in 2009 merges the orders Loganelliiformes, Katoporiida and Shieliiformes into Thelodontiformes, places families Lanarkiidae and Nikoliviidae into Furcacaudiformes on the basis of scale morphology, and establishes Archipelepidiformes as the basal-most order.

A newer taxonomy based on the work of Nelson, Grande and Wilson 2016 and van der Laan 2016.

- Super Class †Thelodontomorphi Jaekel 1911

    o Class †Thelodonti Kiaer 1932

        ▪ Family †Oeseliidae Märss 2005

        ▪ Order †Archipelepidiformes Wilson & Märss 2009

            ◇ Family †Boothialepididae Märss 1999

            ◇ Family †Archipelepididae Märss ex Soehn et al. 2001

        ▪ Order †Furcacaudiformes Wilson & Caldwell 1998 (Fork-tailed thelodonts)

            ◇ Family †Nikoliviidae Karatajūtė-Talimaa 1978

            ◇ Family †Lanarkiidae Obručhev 1949

            ◇ Family †Pezopallichthyidae Wilson & Caldwell 1998

            ◇ Family †Drepanolepididae Wilson & Marss 2009

            ◇ Family †Barlowodidae Märss, Wilson & Thorsteinsson 2002

            ◇ Family †Apalolepididae Turner 1976

◇ Family †Furcacaudidae Wilson & Caldwell 1998

- Clade Thelodontida Stensiö 1958 non Kiaer 1932 s.l.

    ◇ Family †Talivaliidae Marss, Wilson & Thorsteinsson 2002

    ◇ Family †Longodidae Märss 2006b

    ◇ Family †Helenolepididae Wilson & Märss 2009

    ◇ Order †Sandiviiformes Karatajūtė-Talimaa & Märss 2004

        ☐ Family †Angaralepididae Karatajūtė-Talimaa & Märss 2004

        ☐ Family †Stroinolepididae Karatajūtė-Talimaa & Märss 2002

        ☐ Family †Sandiviidae Karatajūtė-Talimaa & Märss 2004

    ◇ Order †Turiniida Stensiö 1958

        ☐ Family †Turiniidae Obručhev 1964

    ◇ Order †Thelodontiformes

        ☐ Family †Thelodontidae Jordan 1905

    ◇ Order †Loganelliiformes Turner 1991

        ☐ Family †Nunavutiidae Marss, Wilson & Thorsteinsson 2002

        ☐ Family †Loganelliidae Märss, Wilson & Thorsteinsson 2002

    ◇ Order †Phlebolepidiformes Berg 1937 s.s.

        ☐ Family †Phlebolepididae Berg 1937 corrig.

        ☐ Family †Shieliidae Märss, Wilson & Thorsteinsson 2002

        ☐ Family †Katoporodidae Soehn et al. 2001 ex Märss, Wilson & Thorsteinsson 2002

Scales

Left to right: denticles of *Paralogania*, *Shielia taiti*, *Lanarkia horrida*

The bony scales of the thelodont group, as the most abundant form of fossil, are also the best understood - and thus most useful. The scales were formed and shed throughout the organisms' lifetimes, and quickly separated after their death.

Bone - being one of the most resistant materials to the process of fossilisation - often preserves internal detail, which allows the histology and growth of the scales to be studied in detail. The scales consist of a non-growing "crown" composed of dentine, with a sometimes-ornamented enameloid upper surface and an aspidine (acellular bony tissue) base. Its growing base is made of cell-free bone, which sometimes developed anchorage structures to fix it in the side of the fish. Beyond that, there appear to be five types of bone-growth, which may represent five natural groupings within the thelodonts - or a spectrum ranging between the end members, meta- (or ortho-) dentine and mesodentine tissues. Interestingly, each of the five scale morphs appears to resemble the scales of more derived groupings of fish, suggesting that thelodont groups may have been stem groups to succeeding clades of fish.

Scale morphology, alone, has limited value for distinguishing thelodont species. Within each organism, scale shape varies greatly according to body area, with intermediate scale forms appearing between different areas; furthermore, scale morphology may not even be constant within a given body area. To confuse things further, scale morphologies are not unique to specific taxa, and may be indistinguishable on the same area of two different species.

The morphology and histology of the thelodonts provides the main tool for quantifying their diversity and distinguishing between species - although ultimately using such convergent traits is prone to errors. Nonetheless, a framework of three groups has been proposed, based upon scale morphology and histology.

Ecology

Furcacauda heintzae

Most thelodonts were probably deposit feeders, although nektonic (actively swimming) forms were probably filter-feeders.. They are mainly known from open shelf environments, but are also found nearer the shore and in some freshwater settings. The appearance of the same species in fresh- and salt-water settings has led to suggestions that some thelodonts migrated into fresh water, perhaps to spawn. However, the transition from fresh- to salt- water should be observable, as the scales' composition would change to reflect the different environment. This compositional change has not yet been found.

Utility as Biostratigraphic Markers

Thelodont scales are globally widespread during the Silurian and Early Devonian times, becoming restricted in range to Gondwana, until their extinction in the Late Devonian (Frasnian). The morphology of some species diversified rapidly enough for the scales to rival the conodonts in utility as biostratigraphic markers, allowing precise correlation of widely spaced sediments.

Evolutionary Patterns

The first major pattern or group of jawless fish with exoskeletons or plated armour, was the Laurentian group, which existed during the Cambrian-Ordovician time. However, the thelodonts (as well as the conodonts, placoderms, acanthodians, and chondrichthyans) are the second major group which are believed to have emerged in the middle Ordovician and lasted near the Late Devonian period. Due to their similar characteristics and chronological time frame of existence, many believe the thelodonts have Laurentian origins.

Cephalaspidomorphi

Cephalaspidomorphs are a group of jawless fishes named for *Cephalaspis* of the osteostracans. Most biologists regard this taxon as extinct, but the name is sometimes used in the classification of lampreys, because lampreys were once thought to be related to cephalaspids. If lampreys are included, they would extend the known range of the group from the Silurian and Devonian periods to the present day.

Biology and Morphology

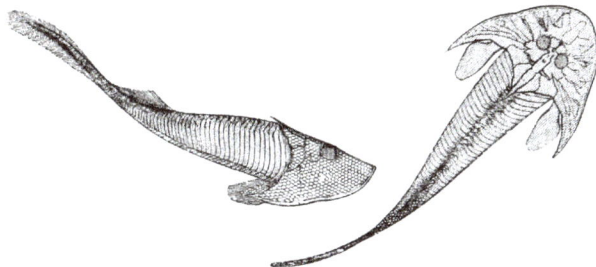

Reconstruction of *Cephalaspis* lyellii

Cephalaspidomorphi were, like most contemporary fishes, very well armoured. The head shield was particularly well developed, protecting the head, gills and the anterior section of the viscera. The body was in most forms well armoured as well. The head shield had a series of grooves over the whole surface, forming an extensive lateral line organ. The eyes were rather small and placed on the top of the head. There was no jaw proper. The mouth opening was surrounded by small plates, making the lips flexible, but without any ability to bite.

No internal skeleton is known, outside of the head shield. If they had a vertebral column at all, it would have been cartilage rather than bone. Likely, the axial skeleton consisted of an unsegmented notochord. A fleshy appendage emerged laterally on each side, behind the head shield, functioning as pectoral fins. The tail had a single, wrap-around tail-fin. Modern fishes with such a tail are rarely quick swimmers, and the cephalaspidomorphi were not likely very active animals. They probably spent much of their time semi-submerged in the mud. They also lacked a swim bladder, and would not have been able to keep afloat without actively swimming. The head shield provided some lift though, and would have made the cephalaspidomorphi better swimmers than most of their contemporaries. The whole group were likely algae- or filter-feeders, combing the bottom for small animals, much like the modern armoured bottom feeders, such as Loricariidae or *Hoplosternum* catfish.

Classification

In the 1920s, the biologists Johan Kiær and Erik Stensiö first recognized the Cephalaspidomorphi as including the osteostracans, anaspids, and lampreys, because all three groups share a single dorsal "nostril", now known as a nasohypophysial opening.

Since then, opinions on the relations among jawless vertebrates have varied. Most workers have come to regard the agnatha as paraphyletic, having given rise to the jawed fishes. Because of shared features such as paired fins, the origins of the jawed vertebrates may lie close to the Cephalaspidomorphi. Many biologists no longer use the name Cephalaspidomorphi because relations among Osteostraci and Anaspida are unclear, and the affinities of the lampreys are also contested. Others have restricted the cephalaspidomorphs to include only groups more clearly related to the Osteostraci, such as Galeaspida and Pituriaspida, that were largely unknown in the 1920s.

Lampreys

Some reference works and databases have regarded Cephalaspidomorphi as a Linnean class whose sole living representatives are the lampreys. Evidence now suggests that lampreys acquired the characters they share with cephalaspids by convergent evolution. As such, many newer works about fishes classify lampreys in a separate group called Petromyzontida or Hyperoartia.

References

- Schwab, IR; Hart, N (2006). "More than black and white". British Journal of Ophthalmology. 90 (4): 406. PMC 1857009. PMID 16572506. doi:10.1136/bjo.2005.085571

- Clack, Jennifer A. (27 June 2012). Gaining Ground, Second Edition: The Origin and Evolution of Tetrapods. Indiana University Press. p. 23. ISBN 0-253-00537-X. Retrieved 12 May 2015

- Laurin, Michel (2 November 2010). How Vertebrates Left the Water. University of California Press. p. 38. ISBN 978-0-520-94798-6. Retrieved 14 May 2015

- Betancur-R; et al. (2013). "The Tree of Life and a New Classification of Bony Fishes.". PLOS Currents Tree of Life (Edition 1). doi:10.1371/currents.tol.53ba26640df0ccaee75bb165c8c26288. Archived from the original on 2013-10-13

- Werren, John; Mart R. Gross; Richard Shine (1980). "Paternity and the evolution of male parentage". Journal of Theoretical Biology. 82 (4). doi:10.1016/0022-5193(80)90182-4. Retrieved 15 September 2013

- Parsons, Alfred Sherwood Romer, Thomas S. (1986). The vertebrate body (6th ed.). Philadelphia: Saunders College Pub. ISBN 978-0-03-910754-3

- Wegner, Nicholas C., Snodgrass, Owen E., Dewar, Heidi, John, Hyde R. Science. "Whole-body endothermy in a mesopelagic fish, the opah, Lampris guttatus". pp. 786–789. Retrieved May 14, 2015

- Thomas J. Near; et al. (2012). "Resolution of ray-finned fish phylogeny and timing of diversification". PNAS. pp. 13698–13703. doi:10.1073/pnas.1206625109

- Laurin, M.; Reisz, R.R. (1995). "A reevaluation of early amniote phylogeny". Zoological Journal of the Linnean Society. 113: 165–223. doi:10.1111/j.1096-3642.1995.tb00932.x

- Romer, Alfred Sherwood; Parsons, Thomas S. (1977). The Vertebrate Body. Philadelphia, PA: Holt-Saunders International. pp. 396–399. ISBN 0-03-910284-X

- Baylis, Jeffrey (1981). "The Evolution of Parental Care in Fishes, with reference to Darwin's rule of male sexual selection". Environmental Biology of Fishes. 6 (2). doi:10.1007/BF00002788. Retrieved 16 September 2013

- Coates, M.I. (2009). "Palaeontology: Beyond the Age of Fishes". Nature. 458: 413–414. PMID 19325614. doi:10.1038/458413a

- Retallack, Gregory (May 2011). "Woodland Hypothesis for Devonian Tetrapod Evolution". Journal of Geology. University of Chicago Press. 119 (3): 235–258. Bibcode:2011JG....119..235R. doi:10.1086/659144

- Kardong, Kenneth (2015). Vertebrates: Comparative Anatomy, Function, Evolution. New York: McGraw-Hill Education. pp. 99–100. ISBN 978-0-07-802302-6

- Grzegorz Niedźwiedzki, Piotr Szrek, Katarzyna Narkiewicz, Marek Narkiewicz & Per E. Ahlberg (2010). "Tetrapod trackways from the early Middle Devonian period of Poland". Nature. Nature Publishing Group. 463 (7277): 43–48. Bibcode:2010Natur.463...43N. PMID 20054388. doi:10.1038/nature08623. Retrieved January 3, 2012

- Avise, J.C.; Mank, J.E. (2009). "Evolutionary perspectives on hermaphroditism in fishes". Sexual Development. 3: 152–163. doi:10.1159/000223079

- Stossel, I. (1995) The discovery of a new Devonian tetrapod trackway in SW Ireland. Journal of the Geological Society, London, 152, 407-413

- Theodore Holmes Bullock; Carl D. Hopkins; Arthur N. Popper (2005). Electroreception. Springer Science+Business Media, Incorporated. p. 229. ISBN 978-0-387-28275-6

- Shanta Barley (January 6, 2010). "Oldest footprints of a four-legged vertebrate discovered". New Scientist. Retrieved January 3, 2010

- Betancur-R, Ricardo; et al. (2013). "The Tree of Life and a New Classification of Bony Fishes". PLOS Currents Tree of Life (Edition 1). doi:10.1371/currents.tol.53ba26640df0ccaee75bb165c8c26288

- Swartz, B. (2012). "A marine stem-tetrapod from the Devonian of Western North America". PLoS ONE. 7 (3): e33683. PMC 3308997. PMID 22448265. doi:10.1371/journal.pone.0033683

- Kardong, Kenneth V. (1998). Vertebrates: Comparative Anatomy, Function, Evolution, second edition, USA: McGraw-Hill, 747 pp.. ISBN 0-07-115356-X/0-697-28654-1

- PLOS. "Ancient southern China fish may have evolved prior to the 'Age of Fish". www.sciencedaily.com. Retrieved 2017-03-11

- Zhu, M; Zhao, W; Jia, L; Lu, J; Qiao, T; Qu, Q (2009). "The oldest articulated osteichthyan reveals mosaic gnathostome characters". Nature. 458: 469–474. PMID 19325627. doi:10.1038/nature07855

- Sigurdsen, Trond; Green, David M. (June 2011). "The origin of modern amphibians: a re-evaluation". Zoological Journal of the Linnean Society. 162 (2): 457–469. ISSN 0024-4082. doi:10.1111/j.1096-3642.2010.00683.x

Understanding the Physiology of Fishes

Fish physiology is the study of the parts and functions of fishes. Some of these parts are jaw, scale, gill and fin. Fish locomotion, electroreception and fish intelligence are some of the significant and important topics related to fish physiology. The following chapter unfolds its crucial aspects in a critical yet systematic manner.

Fish Physiology

When threatened, the toxic pufferfish fills its extremely elastic stomach with water.

Fish physiology is the scientific study of how the component parts of fish function together in the living fish. It can be contrasted with fish anatomy, which is the study of the form or morphology of fishes. In practice, fish anatomy and physiology complement each other, the former dealing with the structure of a fish, its organs or component parts and how they are put together, such as might be observed on the dissecting table or under the microscope, and the later dealing with how those components function together in the living fish.

Respiration

Most fish exchange gases using gills on either side of the pharynx (throat). Gills are tissues which consist of threadlike structures called filaments. These filaments have many functions and "are involved in ion and water transfer as well as oxygen, carbon dioxide, acid and ammonia exchange. Each filament contains a capillary network that provides a large surface area for exchanging oxygen and carbon dioxide. Fish exchange gases by pulling oxygen-rich water through their mouths and pumping it over their gills. In some fish, capillary blood flows in the opposite direction to the water, causing countercurrent exchange. The gills push the oxygen-poor water out through openings in the sides of the pharynx.

Fish from multiple groups can live out of the water for extended time periods. Amphibious fish such as the mudskipper can live and move about on land for up to several days, or live in stagnant or otherwise oxygen depleted water. Many such fish can breathe air via a variety of mechanisms. The skin of anguillid eels may absorb oxygen directly. The buccal cavity of the electric eel may breathe air. Catfish of the families Loricariidae, Callichthyidae, and Scoloplacidae absorb air through their digestive tracts. Lungfish, with the exception of the Australian lungfish, and bichirs have paired lungs similar to those of tetrapods and must surface to gulp fresh air through the mouth and pass spent air out through the gills. Gar and bowfin have a vascularized swim bladder that functions in the same way. Loaches, trahiras, and many catfish breathe by passing air through the gut. Mudskippers breathe by absorbing oxygen across the skin (similar to frogs). A number of fish have evolved so-called accessory breathing organs that extract oxygen from the air. Labyrinth fish (such as gouramis and bettas) have a labyrinth organ above the gills that performs this function. A few other fish have structures resembling labyrinth organs in form and function, most notably snakeheads, pikeheads, and the Clariidae catfish family.

Gills inside the head of a tuna. The head is snout-down, with the view looking towards the mouth. On the right are the detached gills.

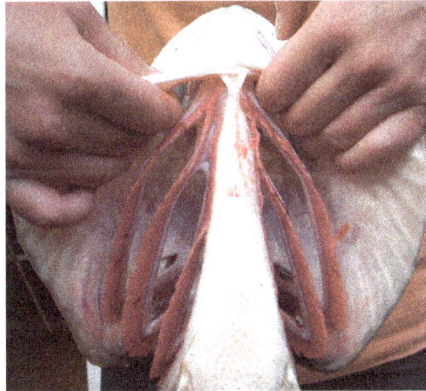

Gill arches bearing gills in a pike

Breathing air is primarily of use to fish that inhabit shallow, seasonally variable waters where the water's oxygen concentration may seasonally decline. Fish dependent solely on dissolved oxygen, such as perch and cichlids, quickly suffocate, while air-breathers survive for much longer, in some cases in water that is little more than wet mud. At the most extreme, some air-breathing fish are able to survive in damp burrows for weeks without water, entering a state of aestivation (summertime hibernation) until water returns.

Air breathing fish can be divided into *obligate* air breathers and *facultative* air breathers. Obligate air breathers, such as the African lungfish, are obligated to breathe air periodically or they suffocate. Facultative air breathers, such as the catfish *Hypostomus plecostomus*, only breathe air if they need to and can otherwise rely on their gills for oxygen. Most air breathing fish are facultative air breathers that avoid the energetic cost of rising to the surface and the fitness cost of exposure to surface predators.

All basal vertebrates breathe with gills. The gills are carried right behind the head, bordering the posterior margins of a series of openings from the esophagus to the exterior. Each gill is supported by a cartilagenous or bony gill arch. The gills of vertebrates typically develop in the walls of the pharynx, along a series of gill slits opening to the exterior. Most species employ a countercurrent exchange system to enhance the diffusion of substances in and out of the gill, with blood and water flowing in opposite directions to each other.

The gills are composed of comb-like filaments, the gill lamellae, which help increase their surface area for oxygen exchange. When a fish breathes, it draws in a mouthful of water at regular intervals. Then it draws the sides of its throat together, forcing the water through the gill openings, so that it passes over the gills to the outside. The bony fish have three pairs of arches, cartilaginous fish have five to seven pairs, while the primitive jawless fish have seven. The vertebrate ancestor no doubt had more arches, as some of their chordate relatives have more than 50 pairs of gills.

Higher vertebrates do not develop gills, the gill arches form during fetal development, and lay the basis of essential structures such as jaws, the thyroid gland, the larynx, the *columella* (corresponding to the stapes in mammals) and in mammals the malleus and incus. Fish gill slits may be the evolutionary ancestors of the tonsils, thymus gland, and Eustachian tubes, as well as many other structures derived from the embryonic branchial pouches.

Scientists have investigated what part of the body is responsible for maintaining the respiratory rhythm. They found that neurons located in the brainstem of fish are responsible for the genesis of the respiratory rhythm. The position of these neurons is slightly different from the centers of respiratory genesis in mammals but they are located in the same brain compartment, which has caused debates about the homology of respiratory centers between aquatic and terrestrial species. In both aquatic and terrestrial respiration, the exact mechanisms by which neurons can generate this involuntary rhythm are still not completely understood.

Another important feature of the respiratory rhythm is that it is modulated to adapt to the oxygen consumption of the body. As observed in mammals, fish "breathe" faster and heavier when they do physical exercise. The mechanisms by which these changes occur have been strongly debated over more than 100 years between scientists. The authors can be classified in 2 schools:

1. Those who think that the major part of the respiratory changes are pre-programmed in the brain, which would imply that neurons from locomotion centers of the brain connect to respiratory centers in anticipation of movements.

2. Those who think that the major part of the respiratory changes result from the detection of muscle contraction, and that respiration is adapted as a consequence of muscular contraction and oxygen consumption. This would imply that the brain possesses some kind of detection mechanisms that would trigger a respiratory response when muscular contraction occurs.

Many now agree that both mechanisms are probably present and complementary, or working alongside a mechanism that can detect changes in oxygen and/or carbon dioxide blood saturation.

Bony Fish

In bony fish, the gills lie in a branchial chamber covered by a bony operculum. The great majority of bony fish species have five pairs of gills, although a few have lost some over the course of evolution. The operculum can be important in adjusting the pressure of water inside of the pharynx to allow proper ventilation of the gills, so that bony fish do not have to rely on ram ventilation (and hence near constant motion) to breathe. Valves inside the mouth keep the water from escaping.

The gill arches of bony fish typically have no septum, so that the gills alone project from the arch, supported by individual gill rays. Some species retain gill rakers. Though all but the most primitive bony fish lack a spiracle, the pseudobranch associated with it often remains, being located at the base of the operculum. This is, however, often greatly reduced, consisting of a small mass of cells without any remaining gill-like structure.

Marine teleosts also use gills to excrete electrolytes. The gills' large surface area tends to create a problem for fish that seek to regulate the osmolarity of their internal fluids. Saltwater is less dilute than these internal fluids, so saltwater fish lose large quantities of water osmotically through their gills. To regain the water, they drink large amounts of seawater and excrete the salt. Freshwater is more dilute than the internal fluids of fish, however, so freshwater fish gain water osmotically through their gills.

In some primitive bony fishes and amphibians, the larvae bear external gills, branching off from the gill arches. These are reduced in adulthood, their function taken over by the gills proper in fishes and by lungs in most amphibians. Some amphibians retain the external larval gills in adulthood, the complex internal gill system as seen in fish apparently being irrevocably lost very early in the evolution of tetrapods.

Cartilaginous Fish

Like other fish, sharks extract oxygen from seawater as it passes over their gills. Unlike other fish, shark gill slits are not covered, but lie in a row behind the head. A modified slit called a spiracle lies just behind the eye, which assists the shark with taking in water during respiration and plays a major role in bottom–dwelling sharks. Spiracles are reduced or missing in active pelagic sharks. While the shark is moving, water passes through the mouth and over the gills in a process known as "ram ventilation". While at rest, most sharks pump water over their gills to ensure a constant supply of oxygenated water. A small number of species have lost the ability to pump water through their gills and must swim without rest. These species are *obligate ram ventilators* and would presumably asphyxiate if unable to move. Obligate ram ventilation is also true of some pelagic bony fish species.

The respiration and circulation process begins when deoxygenated blood travels to the shark's two-chambered heart. Here the shark pumps blood to its gills via the ventral aorta artery where it branches into afferent brachial arteries. Reoxygenation takes place in the gills and the reoxygenated blood flows into the efferent brachial arteries, which come together to form the dorsal aorta. The blood flows from the dorsal aorta throughout the body. The deoxygenated blood from the body then flows through the posterior cardinal veins and enters the posterior cardinal sinuses. From there blood enters the heart ventricle and the cycle repeats.

Sharks and rays typically have five pairs of gill slits that open directly to the outside of the body, though some more primitive sharks have six or seven pairs. Adjacent slits are separated by a cartilaginous gill arch from which projects a long sheet-like septum, partly supported by a further piece of cartilage called the *gill ray*. The individual lamellae of the gills lie on either side of the septum. The base of the arch may also support gill rakers, small projecting elements that help to filter food from the water.

A smaller opening, the spiracle, lies in the back of the first gill slit. This bears a small *pseudobranch* that resembles a gill in structure, but only receives blood already oxygenated by the true gills. The spiracle is thought to be homologous to the ear opening in higher vertebrates.

Most sharks rely on ram ventilation, forcing water into the mouth and over the gills by rapidly swimming forward. In slow-moving or bottom dwelling species, especially among skates and rays, the spiracle may be enlarged, and the fish breathes by sucking water through this opening, instead of through the mouth.

Chimaeras differ from other cartilagenous fish, having lost both the spiracle and the fifth gill slit. The remaining slits are covered by an operculum, developed from the septum of the gill arch in front of the first gill.

Lampreys and Hagfish

Lampreys and hagfish do not have gill slits as such. Instead, the gills are contained in spherical pouches, with a circular opening to the outside. Like the gill slits of higher fish, each pouch contains two gills. In some cases, the openings may be fused together, effectively forming an operculum. Lampreys have seven pairs of pouches, while hagfishes may have six to fourteen, depending on the species. In the hagfish, the pouches connect with the pharynx internally. In adult lampreys, a separate respiratory tube develops beneath the pharynx proper, separating food and water from respiration by closing a valve at its anterior end.

Circulation

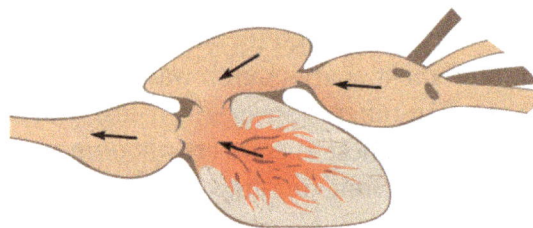

Two-chambered heart of a fish

The circulatory systems of all vertebrates are *closed*, just as in humans. Still, the systems of fish, amphibians, reptiles, and birds show various stages of the evolution of the circulatory system. In fish, the system has only one circuit, with the blood being pumped through the capillaries of the gills and on to the capillaries of the body tissues. This is known as *single cycle* circulation. The heart of fish is therefore only a single pump (consisting of two chambers). Fish have a closed-loop circulatory system. The heart pumps the blood in a single loop throughout the body. In most fish, the heart consists of four parts, including two chambers and an entrance and exit. The first part is the sinus venosus, a thin-walled sac that collects blood from the fish's veins before allowing it to flow to the second part, the atrium, which is a large muscular chamber. The atrium serves as a one-way antechamber, sends blood to the third part, ventricle. The ventricle is another thick-walled, muscular chamber and it pumps the blood, first to the fourth part, bulbus arteriosus, a large tube, and then out of the heart. The bulbus arteriosus connects to the aorta, through which blood flows to the gills for oxygenation.

In amphibians and most reptiles, a double circulatory system is used, but the heart is not always completely separated into two pumps. Amphibians have a three-chambered heart.

Digestion

Jaws allow fish to eat a wide variety of food, including plants and other organisms. Fish ingest food through the mouth and break it down in the esophagus. In the stomach, food is further digested and, in many fish, processed in finger-shaped pouches called pyloric caeca, which secrete digestive enzymes and absorb nutrients. Organs such as the liver and pancreas add enzymes and various chemicals as the food moves through the digestive tract. The intestine completes the process of digestion and nutrient absorption.

In most vertebrates, digestion is a four-stage process involving the main structures of the digestive tract, starting with ingestion, placing food into the mouth, and concluding with the excretion of undigested material through the anus. From the mouth, the food moves to the stomach, where as bolus it is broken down chemically. It then moves to the intestine, where the process of breaking the food down into simple molecules continues and the results are absorbed as nutrients into the circulatory and lymphatic system.

Although the precise shape and size of the stomach varies widely among different vertebrates, the relative positions of the oesophageal and duodenal openings remain relatively constant. As a result, the organ always curves somewhat to the left before curving back to meet the pyloric sphincter. However, lampreys, hagfishes, chimaeras, lungfishes, and some teleost fish have no stomach at all, with the oesophagus opening directly into the intestine. These animals all consume diets that either require little storage of food, or no pre-digestion with gastric juices, or both.

The small intestine is the part of the digestive tract following the stomach and followed by the large intestine, and is where much of the digestion and absorption of food takes place. In fish, the divisions of the small intestine are not clear, and the terms *anterior* or *proximal* intestine may be used instead of duodenum. The small intestine is found in all teleosts, although its form and length vary enormously between species. In teleosts, it is relatively short, typically around one and a half times the length of the fish's body. It commonly has a number of *pyloric caeca*, small pouch-like structures along its length that help to increase the overall surface area of the organ for digesting food. There is no ileocaecal valve in teleosts, with the boundary between the small intestine and the rectum being marked only by the end of the digestive epithelium.

There is no small intestine as such in non-teleost fish, such as sharks, sturgeons, and lungfish. Instead, the digestive part of the gut forms a spiral intestine, connecting the stomach to the rectum. In this type of gut, the intestine itself is relatively straight, but has a long fold running along the inner surface in a spiral fashion, sometimes for dozens of turns. This valve greatly increases both the surface area and the effective length of the intestine. The lining of the spiral intestine is similar to that of the small intestine in teleosts and non-mammalian tetrapods. In lampreys, the spiral valve is extremely small, possibly because their diet requires little digestion. Hagfish have no spiral valve at all, with digestion occurring for almost the entire length of the intestine, which is not subdivided into different regions.

The large intestine is the last part of the digestive system normally found in vertebrate animals. Its function is to absorb water from the remaining indigestible food matter, and then to pass useless waste material from the body. In fish, there is no true large intestine, but simply a short rectum connecting the end of the digestive part of the gut to the cloaca. In sharks, this includes a *rectal gland* that secretes salt to help the animal maintain osmotic balance with the seawater. The gland somewhat resembles a caecum in structure, but is not a homologous structure.

As with many aquatic animals, most fish release their nitrogenous wastes as ammonia. Some of the wastes diffuse through the gills. Blood wastes are filtered by the kidneys.

Saltwater fish tend to lose water because of osmosis. Their kidneys return water to the body. The reverse happens in freshwater fish: they tend to gain water osmotically. Their kidneys produce dilute urine for excretion. Some fish have specially adapted kidneys that vary in function, allowing them to move from freshwater to saltwater.

In sharks, digestion can take a long time. The food moves from the mouth to a J-shaped stomach, where it is stored and initial digestion occurs. Unwanted items may never get past the stomach, and instead the shark either vomits or turns its stomachs inside out and ejects unwanted items from its mouth. One of the biggest differences between the digestive systems of sharks and mammals is that sharks have much shorter intestines. This short length is achieved by the spiral valve with multiple turns within a single short section instead of a long tube-like intestine. The valve provides a long surface area, requiring food to circulate inside the short gut until fully digested, when remaining waste products pass into the cloaca.

Endocrine system

Regulation of Social Behaviour

Oxytocin is a group of neuropeptides found in most vertebrate. One form of oxytocin functions as a hormone which is associated with human love. In 2012, researchers injected cichlids from the social species *Neolamprologus pulcher*, either with this form of isotocin or with a control saline solution. They found isotocin increased "responsiveness to social information", which suggests "it is a key regulator of social behavior that has evolved and endured since ancient times".

Effects of Pollution

Fish can bioaccumulate pollutants that are discharged into waterways. Estrogenic compounds found in pesticides, birth control, plastics, plants, fungi, bacteria, and synthetic drugs leeched into rivers are affecting the endocrine systems of native species. In Boulder, Colorado, white sucker fish found downstream of a municipal waste water treatment plant exhibit impaired or abnormal sexual development. The fish have been exposed to higher levels of estrogen, and leading to feminized fish. Males display female reproductive organs, and both sexes have reduced fertility, and a higher hatch mortality. The same feminization effect can also be seen in male African clawed frogs when exposed to high levels of atrazine, a widely used pesticide. In the marine ecosystem, organochlorine contaminants like pesticides, herbicides (DDT), and chlordan are accumulating within fish tissue and disrupting their endocrine system. High frequencies of infertility and high levels of organochlorines have been found in bonnethead sharks along the Gulf Coast of Florida. These endocrine-disrupting compounds are similar in structure to naturally occurring hormones in fish. They can modulate hormonal interactions in fish by:

- binding to cellular receptors, causing unpredictable and abnormal cell activity

- blocking receptor sites, inhibiting activity

- promoting the creation of extra receptor sites, amplifying the effects of the hormone or compound

- interacting with naturally occurring hormones, changing their shape and impact

- affecting hormone synthesis or metabolism, causing an improper balance or quantity of hormones

Osmoregulation

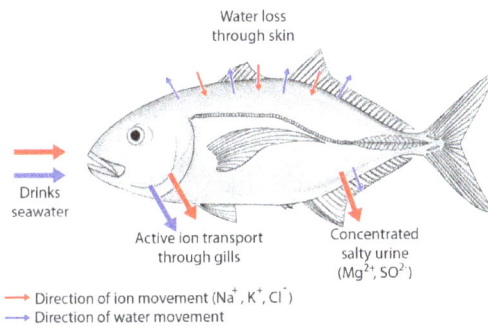

Movement of water and ions in saltwater fish

Two major types of osmoregulation are osmoconformers and osmoregulators. Osmoconformers match their body osmolarity to their environment actively or passively. Most marine invertebrates are osmoconformers, although their ionic composition may be different from that of seawater.

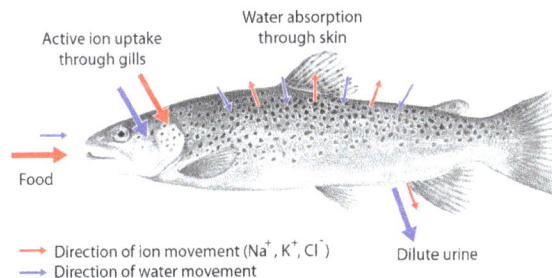

Movement of water and ions in freshwater fish

Osmoregulators tightly regulate their body osmolarity, which always stays constant, and are more common in the animal kingdom. Osmoregulators actively control salt concentrations despite the salt concentrations in the environment. An example is freshwater fish. The gills actively uptake salt from the environment by the use of mitochondria-rich cells. Water will diffuse into the fish, so it excretes a very hypotonic (dilute) urine to expel all the excess water. A marine fish has an internal osmotic concentration lower than that of the surrounding seawater, so it tends to lose water and gain salt. It actively excretes salt out from the gills. Most fish are stenohaline, which means they are restricted to either salt or fresh water and cannot survive in water with a different salt concentration than they are adapted to. However, some fish show a tremendous ability to effectively osmoregulate across a broad range of salinities; fish with this ability are known as euryhaline species, e.g., salmon. Salmon has been observed to inhabit two utterly disparate environments — marine and fresh water — and it is inherent to adapt to both by bringing in behavioral and physiological modifications.

In contrast to bony fish, with the exception of the coelacanth, the blood and other tissue of sharks and Chondrichthyes is generally isotonic to their marine environments because of the high concentration of urea and trimethylamine N-oxide (TMAO), allowing them to be in osmotic balance with the seawater. This adaptation prevents most sharks from surviving in freshwater, and they are therefore confined to marine environments. A few exceptions exist, such as the bull shark, which has developed a way to change its kidney function to excrete large amounts of urea. When a shark dies, the urea is broken down to ammonia by bacteria, causing the dead body to gradually smell strongly of ammonia.

Sharks have adopted a different, efficient mechanism to conserve water, i.e., osmoregulation. They retain urea in their blood in relatively higher concentration. Urea is damaging to living tissue so, to cope with this problem, some fish retain *trimethylamine oxide*. This provides a better solution to urea's toxicity. Sharks, having slightly higher solute concentration (i.e., above 1000 mOsm which is sea solute concentration), do not drink water like fresh water fish.

Thermoregulation

Homeothermy and poikilothermy refer to how stable an organism's temperature is. Most endothermic organisms are homeothermic, like mammals. However, animals with facultative endothermy are often poikilothermic, meaning their temperature can vary considerably. Similarly, most fish are ectotherms, as all of their heat comes from the surrounding water. However, most are homeotherms because their temperature is very stable.

Most organisms have a preferred temperature range, however some can be acclimated to temperatures colder or warmer than what they are typically used to. An organism's preferred temperature is typically the temperature at which the organism's physiological processes can act at optimal rates. When fish become acclimated to other temperatures, the efficiency of their physiological processes may decrease but will continue to function. This is called the thermal neutral zone at which an organism can survive indefinitely.

H. M. Vernon has done work on the death temperature and paralysis temperature (temperature of heat rigor) of various animals. He found that species of the same class showed very similar temperature values, those from the Amphibia examined being 38.5°C, fish 39°C, Reptilia 45°C, and various Molluscs 46°C.

To cope with low temperatures, some fish have developed the ability to remain functional even when the water temperature is below freezing; some use natural antifreeze or antifreeze proteins to resist ice crystal formation in their tissues.

Most sharks are "cold-blooded" or, more precisely, poikilothermic, meaning that their internal body temperature matches that of their ambient environment. Members of the family Lamnidae (such as the shortfin mako shark and the great white shark) are homeothermic and maintain a higher body temperature than the surrounding water. In these sharks, a strip of aerobic red muscle located near the center of the body generates the heat, which the body retains via a countercurrent exchange mechanism by a system of blood vessels called the rete mirabile ("miraculous net"). The common thresher shark has a similar mechanism for maintaining an elevated body temperature, which is thought to have evolved independently.

Tuna can maintain the temperature of certain parts of their body above the temperature of ambient seawater. For example, bluefin tuna maintain a core body temperature of 25–33°C (77–91°F), in water as cold as 6°C (43°F). However, unlike typical endothermic creatures such as mammals and birds, tuna do not maintain temperature within a relatively narrow range. Tuna achieve endothermy by conserving the heat generated through normal metabolism. The rete mirabile ("wonderful net"), the intertwining of veins and arteries in the body's periphery, transfers heat from venous blood to arterial blood via a counter-current exchange system, thus mitigating the effects of surface cooling. This allows the tuna to elevate the temperatures of the highly aerobic tissues of the skeletal muscles, eyes and brain, which supports faster swimming speeds and reduced energy expenditure, and which enables them to survive in cooler waters over a wider range of ocean environments than those of other fish. In all tunas, however, the heart operates at ambient temperature, as it receives cooled blood, and coronary circulation is directly from the gills.

- Homeothermy: Although most fish are exclusively ectothermic, there are exceptions. Certain species of fish maintain elevated body temperatures. Endothermic teleosts (bony fish) are all in the suborder Scombroidei and include the billfishes, tunas, including a "primitive" mackerel species, *Gasterochisma melampus*. All sharks in the family Lamnidae – shortfin mako, long fin mako, white, porbeagle, and salmon shark – are endothermic, and evidence suggests the trait exists in family Alopiidae (thresher sharks). The degree of endothermy varies from the billfish, which warm only their eyes and brain, to bluefin tuna and porbeagle sharks who maintain body temperatures elevated in excess of 20°C above ambient water temperatures. Endothermy, though metabolically costly, is thought to provide advantages such as increased muscle strength, higher rates of central nervous system processing, and higher rates of digestion.

In some fish, a rete mirabile allows for an increase in muscle temperature in regions where this network of vein and arteries is found. The fish is able to thermoregulate certain areas of their body. Additionally, this increase in temperature leads to an increase in basal metabolic temperature. The fish is now able to split ATP at a higher rate and ultimately can swim faster.

The eye of a swordfish can generate heat to better cope with detecting their prey at depths of 2000 feet.

Muscular System

Fish swim by contracting longitudinal red muscle and obliquely oriented white muscles. The red muscle is aerobic and needs oxygen which is supplied by myoglobin. The white muscle is anaerobic and it does not need oxygen. Red muscles are used for sustained activity such as cruising at slow speeds on ocean migrations. White muscles are used for bursts of activity, such as jumping or sudden bursts of speed for catching prey.

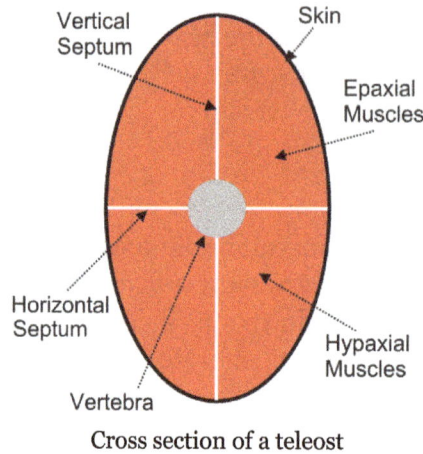

Cross section of a teleost

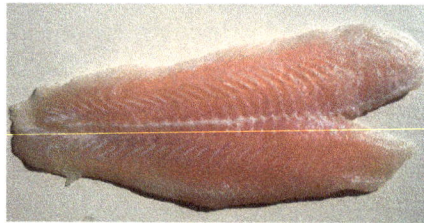

Iridescent shark filets showing myomere structure

Mostly fish have white muscles, but the muscles of some fishes, such as scombroids and salmonids, range from pink to dark red. The red myotomal muscles derive their colour from myoglobin, an oxygen-binding molecule, which tuna express in quantities far higher than most other fish. The oxygen-rich blood further enables energy delivery to their muscles.

Most fish move by alternately contracting paired sets of muscles on either side of the backbone. These contractions form S-shaped curves that move down the body. As each curve reaches the back fin, backward force is applied to the water, and in conjunction with the fins, moves the fish forward. The fish's fins function like an airplane's flaps. Fins also increase the tail's surface area, increasing speed. The streamlined body of the fish decreases the amount of friction from the water.

A typical characteristic of many animals that utilize undulatory locomotion is that they have segmented muscles, or blocks of myomeres, running from their head to tails which are separated by connective tissue called myosepta. In addition, some segmented muscle groups, such the lateral hypaxial musculature in the salamander are oriented at an angle to the longitudinal direction. For these obliquely oriented fibers the strain in the longitudinal direction is greater than the strain in the muscle fiber direction leading to an architectural gear ratio greater than 1. A higher initial angle of orientation and more dorsoventral bulging produces a faster muscle contraction but results in a lower amount of force production. It is hypothesized that animals employ a variable gearing mechanism that allows self-regulation of force and velocity to meet the mechanical demands of the contraction. When a pennate muscle is subjected to a low force, resistance to width changes in the muscle cause it to rotate which consequently produce a higher architectural gear ratio (AGR) (high velocity). However, when subject to a high force, the perpendicular fiber force component overcomes the resistance to width changes and the muscle compresses producing a lower AGR (capable of maintaining a higher force output).

Most fishes bend as a simple, homogenous beam during swimming via contractions of longitudinal red muscle fibers and obliquely oriented white muscle fibers within the segmented axial musculature. The fiber strain (εf) experienced by the longitudinal red muscle fibers is equivalent to the longitudinal strain (εx). The deeper white muscle fibers fishes show diversity in arrangement. These fibers are organized into cone-shaped structures and attach to connective tissue sheets known as myosepta; each fiber shows a characteristic dorsoventral (α) and mediolateral (φ) trajectory. The segmented architecture theory predicts that, $\varepsilon x > \varepsilon f$. This phenomenon results in an architectural gear ratio, determined as longitudinal strain divided by fiber strain ($\varepsilon x / \varepsilon f$), greater than one and longitudinal velocity amplification; furthermore, this emergent velocity amplification may be augmented by variable architectural gearing via mesolateral and dorsoventral shape changes, a pattern seen in pennate muscle contractions. A red-to-white gearing ratio (red εf / white εf) captures the combined effect of the longitudinal red muscle fiber and oblique white muscle fiber strains.

Buoyancy

Swim bladder of a common rudd

The body of a fish is denser than water, so fish must compensate for the difference or they will sink. Many bony fishes have an internal organ called a swim bladder, or gas bladder, that adjusts their buoyancy through manipulation of gases. In this way, fish can stay at the current water depth, or ascend or descend without having to waste energy in swimming. The bladder is only found in bony fishes. In the more primitive groups like some minnows, bichirs and lungfish, the bladder is open to the esophagus and double as a lung. It is often absent in fast swimming fishes such as the tuna and mackerel families. The condition of a bladder open to the esophagus is called physostome, the closed condition physoclist. In the latter, the gas content of the bladder is controlled through the rete mirabilis, a network of blood vessels effecting gas exchange between the bladder and the blood.

Sharks, like this three tonne great white shark, don't have swim bladders. Most sharks need to keep swimming to avoid sinking.

In some fish, a rete mirabile fills the swim bladder with oxygen. A countercurrent exchange system is utilized between the venous and arterial capillaries. By lowering the pH levels in the venous capillaries, oxygen unbinds from blood hemoglobin. This causes an increase in venous blood oxygen concentration, allowing the oxygen to diffuse through the capillary membrane and into the arterial capillaries, where oxygen is still sequestered to hemoglobin. The cycle of diffusion continues until the concentration of oxygen in the arterial capillaries is supersaturated (larger than the concentration of oxygen in the swim bladder). At this point, the free oxygen in the arterial capillaries diffuses into the swim bladder via the gas gland.

Unlike bony fish, sharks do not have gas-filled swim bladders for buoyancy. Instead, sharks rely on a large liver filled with oil that contains squalene, and their cartilage, which is about half the normal density of bone. Their liver constitutes up to 30% of their total body mass. The liver's effectiveness is limited, so sharks employ dynamic lift to maintain depth when not swimming. Sand tiger sharks store air in their stomachs, using it as a form of swim bladder. Most sharks need to constantly swim in order to breathe and cannot sleep very long without sinking (if at all). However, certain species, like the nurse shark, are capable of pumping water across their gills, allowing them to rest on the ocean bottom.

Sensory Systems

Most fish possess highly developed sense organs. Nearly all daylight fish have color vision that is at least as good as a human's (see vision in fishes). Many fish also have chemoreceptors that are responsible for extraordinary senses of taste and smell. Although they have ears, many fish may not hear very well. Most fish have sensitive receptors that form the lateral line system, which detects gentle currents and vibrations, and senses the motion of nearby fish and prey. Sharks can sense frequencies in the range of 25 to 50 Hz through their lateral line.

Fish orient themselves using landmarks and may use mental maps based on multiple landmarks or symbols. Fish behavior in mazes reveals that they possess spatial memory and visual discrimination.

Vision

Vision is an important sensory system for most species of fish. Fish eyes are similar to those of terrestrial vertebrates like birds and mammals, but have a more spherical lens. Their retinas generally have both rod cells and cone cells (for scotopic and photopic vision), and most species have colour vision. Some fish can see ultraviolet and some can see polarized light. Amongst jawless fish, the lamprey has well-developed eyes, while the hagfish has only primitive eyespots. Fish vision shows adaptation to their visual environment, for example deep sea fishes have eyes suited to the dark environment.

Hearing

Hearing is an important sensory system for most species of fish. Hearing threshold and the ability to localize sound sources are reduced underwater, in which the speed of sound is faster than in air. Underwater hearing is by bone conduction, and localization of sound appears to depend on differences in amplitude detected by bone conduction. Aquatic animals such as fish, however, have a more specialized hearing apparatus that is effective underwater.

Fish can sense sound through their lateral lines and their otoliths (ears). Some fishes, such as some species of carp and herring, hear through their swim bladders, which function rather like a hearing aid.

Hearing is well-developed in carp, which have the Weberian organ, three specialized vertebral processes that transfer vibrations in the swim bladder to the inner ear.

Although it is hard to test sharks' hearing, they may have a sharp sense of hearing and can possibly hear prey many miles away. A small opening on each side of their heads (not the spiracle) leads directly into the inner ear through a thin channel. The lateral line shows a similar arrangement, and is open to the environment via a series of openings called lateral line pores. This is a reminder of the common origin of these two vibration- and sound-detecting organs that are grouped together as the acoustico-lateralis system. In bony fish and tetrapods the external opening into the inner ear has been lost.

Chemoreception

The shape of the hammerhead shark's head may enhance olfaction by spacing the nostrils further apart.

Sharks have keen olfactory senses, located in the short duct (which is not fused, unlike bony fish) between the anterior and posterior nasal openings, with some species able to detect as little as one part per million of blood in seawater.

Sharks have the ability to determine the direction of a given scent based on the timing of scent detection in each nostril. This is similar to the method mammals use to determine direction of sound.

They are more attracted to the chemicals found in the intestines of many species, and as a result often linger near or in sewage outfalls. Some species, such as nurse sharks, have external barbels that greatly increase their ability to sense prey.

Magnetoception

Electroreception

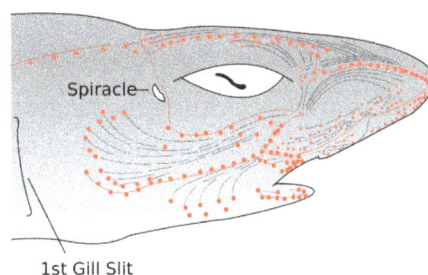

Electromagnetic field receptors (ampullae of Lorenzini) and motion detecting canals in the head of a shark

Some fish, such as catfish and sharks, have organs that detect weak electric currents on the order of millivolt. Other fish, like the South American electric fishes Gymnotiformes, can produce weak electric currents, which they use in navigation and social communication. In sharks, the ampullae of Lorenzini are electroreceptor organs. They number in the hundreds to thousands. Sharks use the ampullae of Lorenzini to detect the electromagnetic fields that all living things produce. This helps sharks (particularly the hammerhead shark) find prey. The shark has the greatest electrical sensitivity of any animal. Sharks find prey hidden in sand by detecting the electric fields they produce. Ocean currents moving in the magnetic field of the Earth also generate electric fields that sharks can use for orientation and possibly navigation.

- The ampullae of Lorenzini allow sharks to sense electrical discharges.

- Electric fish are able to produce electric fields by modified muscles in their body.

Pain

Experiments done by William Tavolga provide evidence that fish have pain and fear responses. For instance, in Tavolga's experiments, toadfish grunted when electrically shocked and over time they came to grunt at the mere sight of an electrode.

In 2003, Scottish scientists at the University of Edinburgh and the Roslin Institute concluded that rainbow trout exhibit behaviors often associated with pain in other animals. Bee venom and acetic acid injected into the lips resulted in fish rocking their bodies and rubbing their lips along the sides and floors of their tanks, which the researchers concluded were attempts to relieve pain, similar to what mammals would do. Neurons fired in a pattern resembling human neuronal patterns.

Professor James D. Rose of the University of Wyoming claimed the study was flawed since it did not provide proof that fish possess "conscious awareness, particularly a kind of awareness that is meaningfully like ours". Rose argues that since fish brains are so different from human brains, fish are probably not conscious in the manner humans are, so that reactions similar to human reactions to pain instead have other causes. Rose had published a study a year earlier arguing that fish cannot feel pain because their brains lack a neocortex. However, animal behaviorist Temple Grandin argues that fish could still have consciousness without a neocortex because "different species can use different brain structures and systems to handle the same functions."

Animal welfare advocates raise concerns about the possible suffering of fish caused by angling. Some countries, such as Germany have banned specific types of fishing, and the British RSPCA now formally prosecutes individuals who are cruel to fish.

Reproductive Processes

Oogonia development in teleosts fish varies according to the group, and the determination of oogenesis dynamics allows the understanding of maturation and fertilisation processes. Changes in the nucleus, ooplasm, and the surrounding layers characterize the oocyte maturation process.

Postovulatory follicles are structures formed after oocyte release; they do not have endocrine function, present a wide irregular lumen, and are rapidly reabsorbed in a process involving the apoptosis of follicular cells. A degenerative process called follicular atresia reabsorbs vitellogenic oocytes not spawned. This process can also occur, but less frequently, in oocytes in other development stages.

Some fish are hermaphrodites, having both testes and ovaries either at different phases in their life cycle or, as in hamlets, have them simultaneously.

Over 97% of all known fish are oviparous, that is, the eggs develop outside the mother's body. Examples of oviparous fish include salmon, goldfish, cichlids, tuna, and eels. In the majority of these species, fertilisation takes place outside the mother's body, with the male and female fish shedding their gametes into the surrounding water. However, a few oviparous fish practice internal fertilisation, with the male using some sort of intromittent organ to deliver sperm into the genital opening of the female, most notably the oviparous sharks, such as the horn shark, and oviparous rays, such as skates. In these cases, the male is equipped with a pair of modified pelvic fins known as claspers.

Marine fish can produce high numbers of eggs which are often released into the open water column. The eggs have an average diameter of 1 millimetre (0.039 in). The eggs are generally surrounded by the extraembryonic membranes but do not develop a shell, hard or soft, around these membranes. Some fish have thick, leathery coats, especially if they must withstand physical force or desiccation. These type of eggs can also be very small and fragile.

| Egg of lamprey | Egg of catshark (mermaids' purse) | Egg of bullhead shark | Egg of chimaera |

The newly hatched young of oviparous fish are called larvae. They are usually poorly formed, carry a large yolk sac (for nourishment) and are very different in appearance from juvenile and adult specimens. The larval period in oviparous fish is relatively short (usually only several weeks), and larvae rapidly grow and change appearance and structure (a process termed metamorphosis) to become juveniles. During this transition larvae must switch from their yolk sac to feeding on zooplankton prey, a process which depends on typically inadequate zooplankton density, starving many larvae.

In ovoviviparous fish the eggs develop inside the mother's body after internal fertilisation but receive little or no nourishment directly from the mother, depending instead on the yolk. Each embryo develops in its own egg. Familiar examples of ovoviviparous fish include guppies, angel sharks, and coelacanths.

Some species of fish are viviparous. In such species the mother retains the eggs and nourishes the embryos. Typically, viviparous fish have a structure analogous to the placenta seen in mammals connecting the mother's blood supply with that of the embryo. Examples of viviparous fish include the surf-perches, splitfins, and lemon shark. Some viviparous fish exhibit oophagy, in which the developing embryos eat other eggs produced by the mother. This has been observed primarily among sharks, such as the shortfin mako and porbeagle, but is known for a few bony fish as well, such as the halfbeak *Nomorhamphus ebrardtii*. Intrauterine cannibalism is an even more unusual mode of vivipary, in which the largest embryos eat weaker and smaller siblings. This behavior is also most commonly found among sharks, such as the grey nurse shark, but has also been reported for *Nomorhamphus ebrardtii*.

In many species of fish, fins have been modified to allow Internal fertilisation.

Aquarists commonly refer to ovoviviparous and viviparous fish as livebearers.

- Many fish species are hermaphrodites. *Synchronous hermaphrodites* possess both ovaries and testes at the same time. *Sequential hermaphrodites* have both types of tissue in their gonads, with one type being predominant while the fish belongs to the corresponding gender.

Fish Jaw

Most bony fishes have two sets of jaws made mainly of bone. The primary oral jaws open and close the mouth, and a second set of pharyngeal jaws are positioned at the back of the throat. The oral jaws are used to capture and manipulate prey by biting and crushing. The pharyngeal jaws, so-called because they are positioned within the pharynx, are used to further process the food and move it from the mouth to the stomach.

Cartilaginous fishes, such as sharks and rays, have one set of oral jaws made mainly of cartilage. They do not have pharyngeal jaws. Generally jaws are articulated and oppose vertically, comprising an upper jaw and a lower jaw and can bear numerous ordered teeth. Bony fishes usually develop only one set of teeth *(monophyodont)*. Cartilaginous fishes grow multiple sets *(polyphyodont)* and replace teeth as they wear.

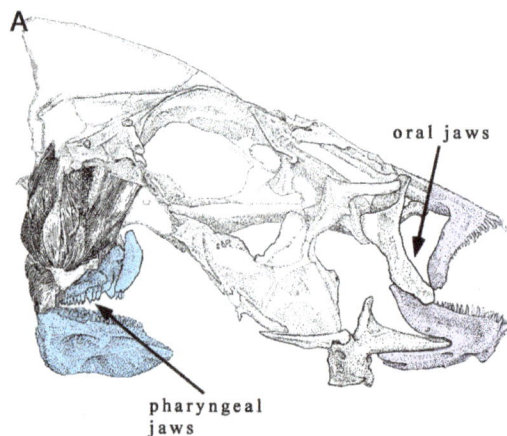

Skull of a generalized cichlid, showing a lateral view of the oral jaws (purple) and the pharyngeal jaws (blue)

Jaws probably originated in the pharyngeal arches supporting the gills of jawless fish. The earliest jaws appeared in the now extinct placoderms and spiny sharks during the Silurian, about 430 million years ago. The original selective advantage offered by the jaw was probably not related to feeding, but to increased respiration efficiency — the jaws were used in the buccal pump to pump water across the gills. The familiar use of jaws for feeding would then have developed as a secondary function before becoming the primary function in many vertebrates. All vertebrate jaws, including the human jaw, evolved from early fish jaws. The appearance of the early vertebrate jaw has been described as "perhaps the most profound and radical evolutionary step in the vertebrate history". Fish without jaws had more difficulty surviving than fish with jaws, and most jawless fish became extinct.

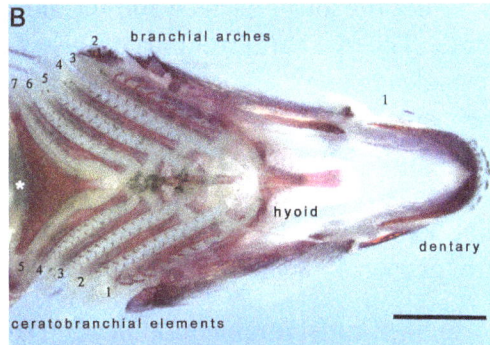

Dorsal view of the lower pharyngeal and oral jaws of a juvenile Malawi eyebiter showing the branchial (pharyngeal) arches and ceratobrachial elements (arch bones). The white asterisk indicates the toothed pharyngeal jaw. Scale bar represents 500 μm.

Jaws use linkage mechanisms. These linkages can be especially common and complex in the head of bony fishes, such as wrasses, which have evolved many specialized feeding mechanisms. Especially advanced are the linkage mechanisms of jaw protrusion. For suction feeding a system of linked four-bar linkages is responsible for the coordinated opening of the mouth and the three-dimensional expansion of the buccal cavity. Other linkages are responsible for protrusion of the premaxilla. Linkage systems are widely distributed in animals. The most thorough overview of the different types of linkages in animals has been provided by M. Muller, who also designed a new classification system, which is especially well suited for biological systems.

Skull

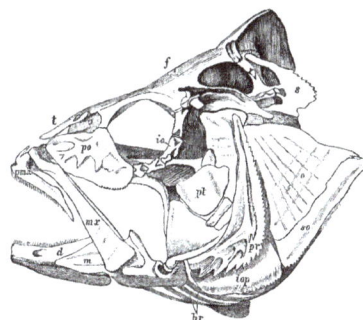

Skeleton of Head of a Perch.

f, frontal.	*pt*, posttympanic.
t, turbinal.	*s*, suprascapula.
po, preorbital.	*o*, opercle.
io, infraorbital ring.	*so*, subopercle.
mx, maxillary.	*pr*, preopercle.
pmx, premaxillary.	*iop*, interopercle.
m, mandible.	*br*, branchiostegal rays.
d, dentary bone.	

The skull of fishes is formed from a series of loosely connected bones. Lampreys and sharks only possess a cartilaginous endocranium, with both the upper and lower jaws being separate elements.

Bony fishes have additional dermal bone, forming a more or less coherent skull roof in lungfish and holost fish. The lower jaw defines a chin.

The simpler structure is found in jawless fish, in which the cranium is represented by a trough-like basket of cartilaginous elements only partially enclosing the brain, and associated with the capsules for the inner ears and the single nostril. Distinctively, these fish have no jaws.

Cartilaginous fish, such as sharks, also have simple skulls. The cranium is a single structure forming a case around the brain, enclosing the lower surface and the sides, but always at least partially open at the top as a large fontanelle. The most anterior part of the cranium includes a forward plate of cartilage, the rostrum, and capsules to enclose the olfactory organs. Behind these are the orbits, and then an additional pair of capsules enclosing the structure of the inner ear. Finally, the skull tapers towards the rear, where the foramen magnum lies immediately above a single condyle, articulating with the first vertebra. There are, in addition, at various points throughout the cranium, smaller foramina for the cranial nerves. The jaws consist of separate hoops of cartilage, almost always distinct from the cranium proper.

In ray-finned fishes, there has also been considerable modification from the primitive pattern. The roof of the skull is generally well formed, and although the exact relationship of its bones to those of tetrapods is unclear, they are usually given similar names for convenience. Other elements of the skull, however, may be reduced; there is little cheek region behind the enlarged orbits, and little, if any bone in between them. The upper jaw is often formed largely from the premaxilla, with the maxilla itself located further back, and an additional bone, the symplectic, linking the jaw to the rest of the cranium.

Although the skulls of fossil lobe-finned fish resemble those of the early tetrapods, the same cannot be said of those of the living lungfishes. The skull roof is not fully formed, and consists of multiple, somewhat irregularly shaped bones with no direct relationship to those of tetrapods. The upper jaw is formed from the pterygoids and vomers alone, all of which bear teeth. Much of the skull is formed from cartilage, and its overall structure is reduced.

Oral Jaws

Lower

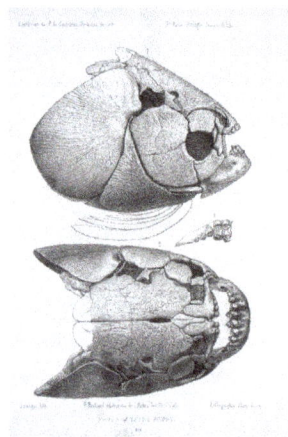

Oral jaw from side and above of *Piaractus brachypomus*, a close relative of piranhas

In vertebrates, the lower jaw (mandible or jawbone) is a bone forming the skull with the cranium. In lobe-finned fishes and the early fossil tetrapods, the bone homologous to the mandible of mammals is merely the largest of several bones in the lower jaw. It is referred to as the *dentary bone*, and forms the body of the outer surface of the jaw. It is bordered below by a number of splenial bones, while the angle of the jaw is formed by a lower angular bone and a suprangular bone just above it. The inner surface of the jaw is lined by a *prearticular* bone, while the articular bone forms the articulation with the skull proper. Finally a set of three narrow *coronoid bones* lie above the prearticular bone. As the name implies, the majority of the teeth are attached to the dentary, but there are commonly also teeth on the coronoid bones, and sometimes on the prearticular as well.

This complex primitive pattern has, however, been simplified to various degrees in the great majority of vertebrates, as bones have either fused or vanished entirely. In teleosts, only the dentary, articular, and angular bones remain. Cartilagenous fish, such as sharks, do not have any of the bones found in the lower jaw of other vertebrates. Instead, their lower jaw is composed of a cartilagenous structure homologous with the Meckel's cartilage of other groups. This also remains a significant element of the jaw in some primitive bony fish, such as sturgeons.

Upper

The upper jaw, or maxilla is a fusion of two bones along the palatal fissure that form the upper jaw. This is similar to the mandible (lower jaw), which is also a fusion of two halves at the mandibular symphysis. In bony fish, the maxilla is called the "upper maxilla," with the mandible being the "lower maxilla". The alveolar process of the maxilla holds the upper teeth, and is referred to as the maxillary arch. In most vertebrates, the foremost part of the upper jaw, to which the incisors are attached in mammals consists of a separate pair of bones, the premaxillae. In bony fish, both maxilla and premaxilla are relatively plate-like bones, forming only the sides of the upper jaw, and part of the face, with the premaxilla also forming the lower boundary of the nostrils. Cartilaginous fish, such as sharks and rays also lack a true maxilla. Their upper jaw is instead formed from a cartilagenous bar that is not homologous with the bone found in other vertebrates.

Some fish have permanently protruding upper jawbones called rostrums. Billfish (marlin, swordfish and sailfish) use rostrums (bills) to slash and stun prey. Paddlefish, goblin sharks and hammerhead sharks have rostrums packed with electroreceptors which signal the presence of prey by detecting weak electrical fields. Sawsharks and the critically endangered sawfish have rostrums (saws) which are both electro-sensitive and used for slashing. The rostrums extend ventrally in front of the fish. In the case of hammerheads the rostrum (hammer) extends both ventrally and laterally (sideways).

Fish with rostrums (extended upper jawbones)

Sailfish, like all billfish, have a rostrum (bill) which evolved from the upper jawbone

The paddlefish has a rostrum packed with electroreceptors

Sawfish have an electro-sensitive rostrum (saw) which is also used to slash at prey

Pharyngeal Jaws

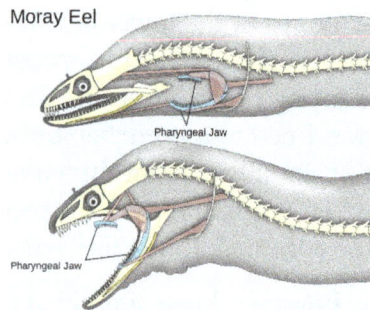

Moray eels have two sets of jaws: the oral jaws that capture prey and the pharyngeal jaws that advance into the mouth and move prey from the oral jaws to the esophagus for swallowing

Pharyngeal jaws are a second set of jaws distinct from the primary (oral) jaws. They are contained within the throat, or pharynx, of most bony fish. They are believed to have originated, in a similar way to oral jaws, as a modification of the fifth gill arch which no longer has a respiratory function. The first four arches still function as gills. Unlike the oral jaw, the pharyngeal jaw has no jaw joint, but is supported instead by a sling of muscles.

Pharyngeal jaw of an asp carrying some pharyngeal teeth

A notable example occurs with the moray eel. The pharyngeal jaws of most fishes are not mobile. The pharyngeal jaws of the moray are highly mobile, perhaps as an adaptation to the constricted nature of the burrows they inhabit which inhibits their ability to swallow as other fishes do by creating a negative pressure in the mouth. Instead, when the moray bites prey, it first bites normally with its oral jaws, capturing the prey. Immediately thereafter, the pharyngeal jaws are brought forward and bite down on the prey to grip it; they then retract, pulling the prey down the moray eel's gullet, allowing it to be swallowed.

All vertebrates have a pharynx, used in both feeding and respiration. The pharynx arises during development through a series of six or more outpocketings called pharyngeal arches on the lateral sides of the head. The pharyngeal arches give rise to a number of different structures in the skeletal, muscular and circulatory systems in a manner which varies across the vertebrates. Pharyngeal arches trace back through chordates to basal deuterostomes who also share endodermal outpocketings of the pharyngeal apparatus. Similar patterns of gene expression can be detected in the developing pharynx of amphioxus and hemichordates. However, the vertebrate pharynx is unique in that it gives rise to endoskeletal support through the contribution of neural crest cells.

Cartilaginous Jaws

Cartilaginous fishes (sharks, rays and skates) have cartilaginous jaws. The jaw's surface (in comparison to the vertebrae and gill arches) needs extra strength due to its heavy exposure to physical stress. It has a layer of tiny hexagonal plates called "tesserae", which are crystal blocks of calcium salts arranged as a mosaic. This gives these areas much of the same strength found in the bony tissue found in other animals.

Generally sharks have only one layer of tesserae, but the jaws of large specimens, such as the bull shark, tiger shark, and the great white shark, have two to three layers or more, depending on body size. The jaws of a large great white shark may have up to five layers. In the rostrum (snout), the cartilage can be spongy and flexible to absorb the power of impacts.

In sharks and other extant elasmobranchs the upper jaw is not fused to the cranium, and the lower jaw is articulated with the upper. The arrangement of soft tissue and any additional articulations connecting these elements is collectively known as the jaw suspension. There are several archetypal jaw suspensions: amphistyly, orbitostyly, hyostyly, and euhyostyly. In amphistyly, the palatoquadrate has a postorbital articulation with the chondrocranium from which ligaments primarily suspend it anteriorly. The hyoid articulates with the mandibular arch posteriorly, but it appears to provide little support to the upper and lower jaws. In orbitostyly, the orbital process hinges with the orbital wall and the hyoid provides the majority of suspensory support. In contrast, hyostyly involves an ethmoid articulation between the upper jaw and the cranium, while the hyoid most likely provides vastly more jaw support compared to the anterior ligaments. Finally, in euhyostyly, also known as true hyostyly, the mandibular cartilages lack a ligamentous connection to the cranium. Instead, the hyomandibular cartilages provide the only means of jaw support, while the ceratohyal and basihyal elements articulate with the lower jaw, but are disconnected from the rest of the hyoid.

Teeth

Inside of a shark jaw where new teeth move forward as though on a conveyor belt

Jaws provide a platform in most fishes for simple pointed teeth. Lungfish and chimaera have teeth modified into broad enamel plates with jagged ridges for crushing or grinding. Carp and loach have pharyngeal teeth only. Sea horses, pipefish and adult sturgeon have no teeth of any type.

Shark teeth are embedded in the gums rather than directly affixed to the jaw, and are constantly replaced throughout life. Multiple rows of replacement teeth grow in a groove on the inside of the jaw and steadily moving forward as though on a conveyor belt. Some sharks lose 30,000 or more teeth in their lifetime. The rate of tooth replacement varies from once every 8 to 10 days to several months. In most species, teeth are replaced one at a time as opposed to the simultaneous replacement of an entire row, which is observed in the cookiecutter shark.

Tooth shape depends on the shark's diet: those that feed on mollusks and crustaceans have dense and flattened teeth used for crushing, those that feed on fish have needle-like teeth for gripping, and those that feed on larger prey such as mammals have pointed lower teeth for gripping and triangular upper teeth with serrated edges for cutting. The teeth of plankton-feeders such as the basking shark are small and non-functional.

Cartilaginous jaws and their teeth

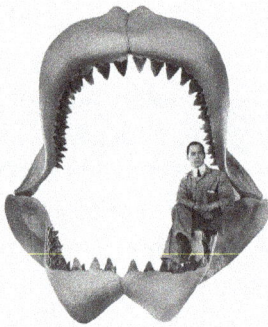

Jaw reconstruction of the extinct *Carcharodon megalodon*, 1909

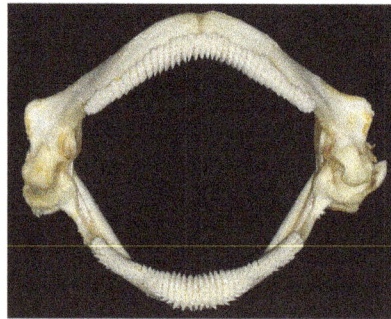

The thornback ray has teeth adapted to feed on crabs, shrimps and small fish.

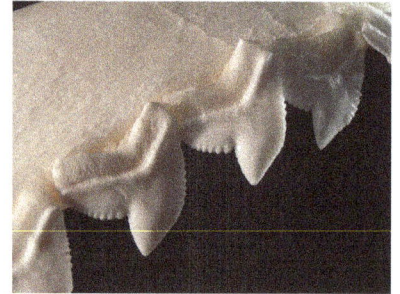

Tiger shark teeth are oblique and serrated to saw through flesh

Examples

Salmon

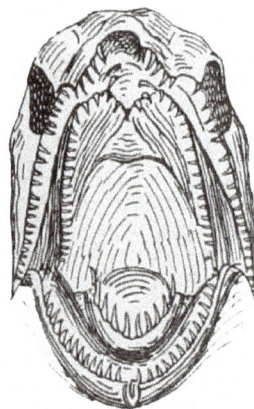

Open mouth of a salmon showing the second set of pharyngeal jaws positioned at the back of the throat

Male salmon often remodel their jaws during spawning runs so they have a pronounced curvature. These hooked jaws are called kypes. The purpose of the kype is not altogether clear, though they can be used to establish dominance by clamping them around the base of the tail (caudal peduncle) of an opponent.

Kype of a spawning male salmon

Cichlids

Fish jaws, like vertebrates in general, normally show bilateral symmetry. An exception occurs with the parasitic scale-eating cichlid *Perissodus microlepis*. The jaws of this fish occur in two distinct morphological forms. One morph has its jaw twisted to the left, allowing it to eat scales more readily on its victim's right flank. The other morph has its jaw twisted to the right, which makes it easier to eat scales on its victim's left flank. The relative abundance of the two morphs in populations is regulated by frequency-dependent selection.

Wrasses

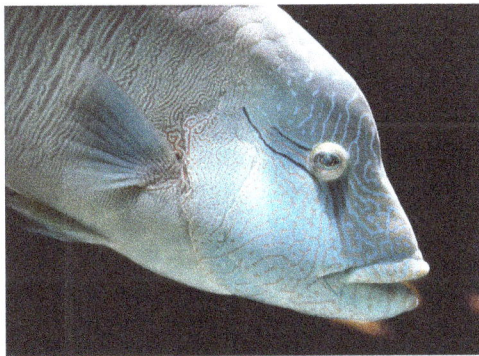
Lips of a humphead wrasse

Wrasses have become a primary study species in fish-feeding biomechanics due to their jaw structure. They have protractile mouths, usually with separate jaw teeth that jut outwards. Many species can be readily recognized by their thick lips, the inside of which is sometimes curiously folded, a peculiarity which gave rise the German name of "lip-fishes" (*Lippfische.*)

The nasal and mandibular bones are connected at their posterior ends to the rigid neurocranium, and the superior and inferior articulations of the maxilla are joined to the anterior tips of these two bones, respectively, creating a loop of 4 rigid bones connected by moving joints. This "four-bar linkage" has the property of allowing numerous arrangements to achieve a given mechanical result (fast jaw protrusion or a forceful bite), thus decoupling morphology from function. The actual morphology of wrasses reflects this, with many lineages displaying different jaw morphology that results in the same functional output in a similar or identical ecological niche.

Other

Relative to its size the stoplight loosejaw has one of the widest gapes of any fish.

Stoplight loosejaws are small fish found worldwide in the deep sea. Relative to their size they have one of the widest gapes of any fish. The lower jaw has no ethmoid membrane (floor) and is attached only by the hinge and a modified tongue bone. There are several large, fang-like teeth in the front of the jaws, followed by many small barbed teeth. There are several groups of pharyngeal teeth that serve to direct food down the esophagus.

Another deep sea fish, the pelican eel, has jaws larger than its body. The jaws are lined with small teeth and are loosely hinged. They open wide enough to swallow a fish larger than the eel itself.

Distichodontidae are a family of fresh water fishes which can be divided into genera with protractile upper jaws which are carnivores, and genera with nonprotractile upper jaws which are herbivores or predators of very small organisms.

Evolution

Vertebrate Classes

The appearance of the early vertebrate jaw has been described as "a crucial innovation" and "perhaps the most profound and radical evolutionary step in the vertebrate history". Fish without jaws had more difficulty surviving than fish with jaws, and most jawless fish became extinct. However studies of the cyclostomes, the jawless hagfishes and lampreys that did survive, have yielded little insight into the deep remodelling of the vertebrate skull that must have taken place as early jaws evolved.

The customary view is that jaws are homologous to the gill arches. In jawless fishes a series of gills opened behind the mouth, and these gills became supported by cartilaginous elements. The first set of these elements surrounded the mouth to form the jaw. The upper portion of the second embryonic arch supporting the gill became the hyomandibular bone of jawed fishes, which supports the skull and therefore links the jaw to the cranium. The hyomandibula is a set of bones found in the hyoid region in most fishes. It usually plays a role in suspending the jaws or the operculum in the case of teleosts.

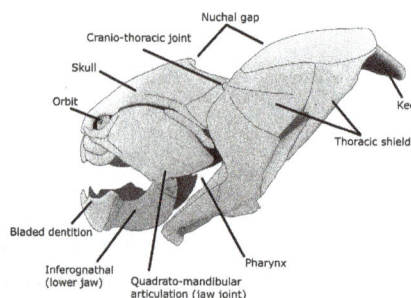

Skull diagram of the huge predatory placoderm fish *Dunkleosteus terrelli*, which lived about 380–360 million years ago

Reconstruction of *Dunkleosteus terrelli*

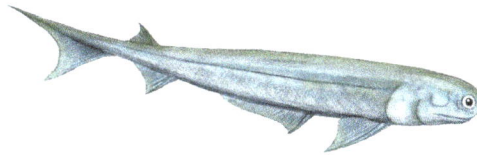

Spiny shark

It is now accepted that the precursors of the jawed vertebrates are the long extinct bony (armoured) jawless fish, the so-called ostracoderms. The earliest known fish with jaws are the now extinct placoderms and spiny sharks.

Placoderms were a class of fish, heavily armoured at the front of their body, which first appeared in the fossil records during the Silurian about 430 million years ago. Initially they very successful, diversifying remarkably during the Devonian. They became extinct by the end of that period, about 360 million years ago. Their largest species, *Dunkleosteus terrelli*, measured up to 10 m (33 ft) and weighed 3.6 t (4.0 short tons). It possessed a four bar linkage mechanism for jaw opening that incorporated connections between the skull, the thoracic shield, the lower jaw and the jaw muscles joined together by movable joints. This mechanism allowed *Dunkleosteus terrelli* to achieve a high speed of jaw opening, opening their jaws in 20 milliseconds and completing the whole process in 50-60 milliseconds, comparable to modern fishes that use suction feeding to assist in prey capture. They could also produce high bite forces when closing the jaw, estimated at 6,000 N (1,350 lb$_f$) at the tip and 7,400 N (1,660 lb$_f$) at the blade edge in the largest individuals. The pressures generated in those regions were high enough to puncture or cut through cuticle or dermal armour suggesting that *Dunkleosteus terrelli* was perfectly adapted to prey on free-swimming, armoured prey like arthropods, ammonites, and other placoderms.

Spiny sharks were another class of fish which appeared also in the fossil records during the Silurian at about the same time as the placoderms. They were smaller than most placoderms, usually under 20 centimetres. Spiny sharks did not diversify as much as placoderms, but survived much longer into the Early Permian about 290 million years ago.

The original selective advantage offered by the jaw was not related to feeding, but to increased respiration efficiency. The jaws were used in the buccal pump still observable in modern fish and amphibians, that uses "breathing with the cheeks" to pump water across the gills of fish or air into the lungs in the case of amphibians. Over evolutionary time the more familiar use of jaws (to humans) for feeding was selected for and became a very important function in vertebrates. Many teleost fish have substantially modified jaws for suction feeding and jaw protrusion, resulting in highly complex jaws with dozens of bones involved.

Jaws are thought to derive from the pharyngeal arches that support the gills in fish. The two most anterior of these arches are thought to have become the jaw itself (see hyomandibula) and the hyoid arch, which braces the jaw against the braincase and increases mechanical efficiency. While there is no fossil evidence directly to support this theory, it makes sense in light of the numbers of pharyngeal arches that are visible in extant jawed (the Gnathostomes), which have seven arches, and primitive jawless vertebrates (the Agnatha), which have nine.

Meckel's cartilage is a piece of cartilage from which the mandibles (lower jaws) of vertebrates evolved. Originally it was the lower of two cartilages which supported the first gill arch (nearest the front) in early fish. Then it grew longer and stronger, and acquired muscles capable of closing the developing jaw. In early fish and in chondrichthyans (cartilaginous fish such as sharks), Meckel's cartilage continued to be the main component of the lower jaw. But in the adult forms of osteichthyans (bony fish) and their descendants (amphibians, reptiles, birds and mammals) the cartilage was covered in bone - although in their embryos the jaw initially develops as the Meckel's cartilage. In tetrapods the cartilage partially ossifies (changes to bone) at the rear end of the jaw and becomes the articular bone, which forms part of the jaw joint in all tetrapods except mammals.

Fish Gill

Most fish exchange gases using gills on either side of the pharynx (throat). Gills are tissues which consist of threadlike protein structures called filaments. These filaments have many functions including the transfer of ions and water, as well as the exchange of oxygen, carbon dioxide, acids and ammonia. Each filament contains a capillary network that provides a large surface area for exchanging oxygen and carbon dioxide. Fish exchange gases by pulling oxygen-rich water through their mouths and pumping it over their gills. In some fish, capillary blood flows in the opposite direction to the water, causing countercurrent exchange. The gills push the oxygen-poor water out through openings in the sides of the pharynx. Some fish, like sharks and lampreys, possess multiple gill openings. However, bony fish have a single gill opening on each side. This opening is hidden beneath a protective bony cover called an operculum.

Juvenile bichirs have external gills, a very primitive feature that they share with larval amphibians.

Breathing with Gills

Air breathing fish can be divided into *obligate* air breathers and *facultative* air breathers. Obligate air breathers, such as the African lungfish, are obligated to breathe air periodically or they suffocate. Facultative air breathers, such as the catfish *Hypostomus plecostomus*, only breathe air if they need to and can otherwise rely on their gills for oxygen. Most air breathing fish are facultative air breathers that avoid the energetic cost of rising to the surface and the fitness cost of exposure to surface predators.

All basal vertebrates breathe with gills. The gills are carried right behind the head, bordering the posterior margins of a series of openings from the esophagus to the exterior. Each gill is supported by a cartilagenous or bony gill arch. The gills of vertebrates typically develop in the walls of the pharynx, along a series of gill slits opening to the exterior. Most species employ a countercurrent exchange system to enhance the diffusion of substances in and out of the gill, with blood and water flowing in opposite directions to each other.

The gills are composed of comb-like filaments, the gill lamellae, which help increase their surface area for oxygen exchange. When a fish breathes, it draws in a mouthful of water at regular intervals. Then it draws the sides of its throat together, forcing the water through the gill openings, so that it passes over the gills to the outside. The bony fish have three pairs of arches, cartilaginous fish have five to seven pairs, while the primitive jawless fish have seven. The vertebrate ancestor no doubt had more arches, as some of their chordate relatives have more than 50 pairs of gills.

Gills usually consist of thin filaments of tissue, branches, or slender tufted processes that have a highly folded surface to increase surface area. The high surface area is crucial to the gas exchange of aquatic organisms as water contains only a small fraction of the dissolved oxygen that air does. A cubic meter of air contains about 250 grams of oxygen at STP. The concentration of oxygen in water is lower than air and it diffuses more slowly. In a litre of freshwater the oxygen content is 8 cm^3 per litre compared to 210 in the same volume of air. Water is 777 times more dense than air and is 100 times more viscous. Oxygen has a diffusion rate in air 10,000 times greater than in water. The use of sac-like lungs to remove oxygen from water would not be efficient enough to sustain life. Rather than using lungs "Gaseous exchange takes place across the surface of highly vascularised gills over which a one-way current of water is kept flowing by a specialised pumping mechanism. The density of the water prevents the gills from collapsing and lying on top of each other, which is what happens when a fish is taken out of water."

Higher vertebrates do not develop gills, the gill arches form during fetal development, and lay the basis of essential structures such as jaws, the thyroid gland, the larynx, the *columella* (corresponding to the stapes in mammals) and in mammals the malleus and incus. Fish gill slits may be the evolutionary ancestors of the tonsils, thymus gland, and Eustachian tubes, as well as many other structures derived from the embryonic branchial pouches.

Bony Fish

In bony fish, the gills lie in a branchial chamber covered by a bony operculum (*branchia* is an Ancient Greek word for gills). The great majority of bony fish species have five pairs of gills, although a few have lost some over the course of evolution. The operculum can be important in adjusting the pressure of water inside of the pharynx to allow proper ventilation of the gills, so that bony fish do not have to rely on ram ventilation (and hence near constant motion) to breathe. Valves inside the mouth keep the water from escaping.

The gill arches of bony fish typically have no septum, so that the gills alone project from the arch, supported by individual gill rays. Some species retain gill rakers. Though all but the most primitive bony fish lack a spiracle, the pseudobranch associated with it often remains, being located at the base of the operculum. This is, however, often greatly reduced, consisting of a small mass of cells without any remaining gill-like structure.

Marine teleosts also use gills to excrete electrolytes. The gills' large surface area tends to create a problem for fish that seek to regulate the osmolarity of their internal fluids. Saltwater is less dilute than these internal fluids, so saltwater fish lose large quantities of water osmotically through their gills. To regain the water, they drink large amounts of seawater and excrete the salt. Freshwater is more dilute than the internal fluids of fish, however, so freshwater fish gain water osmotically through their gills.

In some primitive bony fishes and amphibians, the larvae bear external gills, branching off from the gill arches. These are reduced in adulthood, their function taken over by the gills proper in fishes and by lungs in most amphibians. Some amphibians retain the external larval gills in adulthood, the complex internal gill system as seen in fish apparently being irrevocably lost very early in the evolution of tetrapods.

Cartilaginous Fish

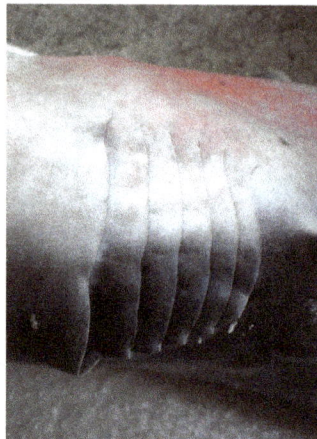

Six gill slits in a bigeyed sixgill shark; most sharks have only five

Sharks and rays typically have five pairs of gill slits that open directly to the outside of the body, though some more primitive sharks have six or seven pairs. Adjacent slits are separated by a cartilaginous gill arch from which projects a long sheet-like septum, partly supported by a further piece of cartilage called the gill ray. The individual lamellae of the gills lie on either side of the septum. The base of the arch may also support gill rakers, small projecting elements that help to filter food from the water.

A smaller opening, the spiracle, lies in the back of the first gill slit. This bears a small pseudobranch that resembles a gill in structure, but only receives blood already oxygenated by the true gills. The spiracle is thought to be homologous to the ear opening in higher vertebrates.

Most sharks rely on ram ventilation, forcing water into the mouth and over the gills by rapidly swimming forward. In slow-moving or bottom dwelling species, especially among skates and rays, the spiracle may be enlarged, and the fish breathes by sucking water through this opening, instead of through the mouth.

Chimaeras differ from other cartilagenous fish, having lost both the spiracle and the fifth gill slit. The remaining slits are covered by an operculum, developed from the septum of the gill arch in front of the first gill.

The shared trait of breathing via gills in bony fish and cartilaginous fish is a famous example of symplesiomorphy. Bony fish are more closely related to terrestrial vertebrates, which evolved out of a clade of bony fishes that breathe through their skin or lungs, than they are to the sharks, rays, and the other cartilaginous fish. Their kind of gill respiration is shared by the "fishes" because it was present in their common ancestor and lost in the other living vertebrates. But based on this shared trait, we cannot infer that bony fish are more closely related to sharks and rays than they are to terrestrial vertebrates.

Lampreys and Hagfish

Lampreys and hagfish do not have gill slits as such. Instead, the gills are contained in spherical pouches, with a circular opening to the outside. Like the gill slits of higher fish, each pouch contains two gills. In some cases, the openings may be fused together, effectively forming an operculum. Lampreys have seven pairs of pouches, while hagfishes may have six to fourteen, depending on the species. In the hagfish, the pouches connect with the pharynx internally. In adult lampreys, a separate respiratory tube develops beneath the pharynx proper, separating food and water from respiration by closing a valve at its anterior end.

Breathing without Gills

Although most fish respire primarily using gills, some fishes can at least partially respire using mechanisms that do not require gills. In some species cutaneous respiration accounts for 5 to 40 percent of the total respiration, depending on temperature. Cutaneous respiration is more important in species that breathe air, such as mudskippers and reedfish, and in such species can account for nearly half the total respiration.

Fish from multiple groups can live out of the water for extended time periods. Amphibious fish such as the mudskipper can live and move about on land for up to several days, or live in stagnant or otherwise oxygen depleted water. Many such fish can breathe air via a variety of mechanisms. The skin of anguillid eels may absorb oxygen directly. The buccal cavity of the electric eel may breathe air. Catfish of the families Loricariidae, Callichthyidae, and Scoloplacidae absorb air through their digestive tracts. Lungfish, with the exception of the Australian lungfish, and bichirs have paired lungs similar to those of tetrapods and must surface to gulp fresh air through the mouth and pass spent air out through the gills. Gar and bowfin have a vascularized swim bladder that functions in the same way. Loaches, trahiras, and many catfish breathe by passing air through the gut. Mudskippers breathe by absorbing oxygen across the skin (similar to frogs). A number of fish have evolved so-called accessory breathing organs that extract oxygen from the air. Labyrinth fish (such as gouramis and bettas) have a labyrinth organ above the gills that performs this function. A few other fish have structures resembling labyrinth organs in form and function, most notably snakeheads, pikeheads, and the Clariidae catfish family.

Breathing air is primarily of use to fish that inhabit shallow, seasonally variable waters where the water's oxygen concentration may seasonally decline. Fish dependent solely on dissolved oxygen, such as perch and cichlids, quickly suffocate, while air-breathers survive for much longer, in some cases in water that is little more than wet mud. At the most extreme, some air-breathing fish are able to survive in damp burrows for weeks without water, entering a state of aestivation (summertime hibernation) until water returns.

Parasites on Gills

Fish gills are the preferred habitat of many ectoparasites (parasites attached to the gill but living out of it); the most commons are monogeneans and certain groups of parasitic copepods, which can be extremely numerous. Other ectoparasites found on gills are leeches and, in seawater, larvae of gnathiid isopods. Endoparasites (parasites living inside the gills) include encysted adult didymozoid trematodes, a few trichosomoidid nematodes of the genus *Huffmanela*, including *Huff-*

manela ossicola which lives within the gill bone, and the encysted parasitic turbellarian *Paravortex*. Various protists and Myxosporea are also parasitic on gills, where they form cysts.

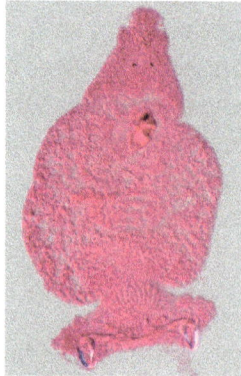
Monogenean parasite on the gill of a grouper

Fish Scale

Cycloid scales cover these teleost fish (rohu)

The skin of most fishes are covered with scales, which, in many cases, are animal reflectors or produce animal coloration. Scales vary enormously in size, shape, structure, and extent, ranging from strong and rigid armour plates in fishes such as shrimpfishes and boxfishes, to microscopic or absent in fishes such as eels and anglerfishes. The morphology of a scale can be used to identify the species of fish it came from.

Cartilaginous fishes (sharks and rays) are covered with placoid scales. Most bony fishes are covered with the cycloid scales of salmon and carp, or the ctenoid scales of perch, or the ganoid scales of sturgeons and gars. Some species are covered instead by scutes, and others have no outer covering on the skin.

Fish scales are part of the fish's integumentary system, and are produced from the mesoderm layer of the dermis, which distinguishes them from reptile scales. The same genes involved in tooth and hair development in mammals are also involved in scale development. The placoid scales of cartilaginous fishes are also called dermal denticles and are structurally homologous with vertebrate teeth. It has been suggested that the scales of bony fishes are similar in structure to teeth, but they probably originate from different tissue. Most fish are also covered in a protective layer of mucus (slime).

Placoid Scales

Cartilaginous fishes, like this tiger shark, have placoid scales (dermal denticles)

Placoid scales are found in the cartilaginous fishes: sharks, rays, and chimaeras. They are also called *dermal denticles*. Placoid scales are structurally homologous with vertebrate teeth ("denticle" translates to "small tooth"), having a central pulp cavity supplied with blood vessels, surrounded by a conical layer of dentine, all of which sits on top of a rectangular basal plate that rests on the dermis. The outermost layer is composed of vitrodentine, a largely inorganic enamel-like substance. Placoid scales cannot grow in size, but rather more scales are added as the fish increases in size.

Similar scales can also be found under the head of the denticle herring. The amount of scale coverage is much less in rays and chimaeras.

The skin of sharks is entirely covered by placoid scales. The scales are supported by spines, which feel rough when stroked in a backward direction, but when flattened by the forward movement of water, create tiny vortices that reduce hydrodynamic drag, making swimming both more efficient as well as quieter compared to that of bony fishes. The rough, sandpaper-like texture of shark and ray skin, coupled with its toughness, has led it to be valued as a source of rawhide leather, called shagreen. One of the many historical applications of shark shagreen was in making hand-grips for swords.

Unlike bony fish, sharks have a complicated dermal corset made of flexible collagenous fibers arranged as a helical network surrounding their body. The corset works as an outer skeleton, providing attachment for their swimming muscles and thus saving energy. Their dermal teeth give them hydrodynamic advantages, as the scales reduce the turbulence of swimming.

Leptoid Scales

Leptoid scales are found on higher-order bony fish, the teleosts (the more derived clade of ray-finned fishes). As the fish grow, scales are added in concentric layers. The scales are arranged so as to overlap in a head-to-tail configuration, like roof tiles, allowing a smoother flow of water over the body and thereby reducing drag. Leptoid scales come in two forms: cycloid and ctenoid.

Cycloid Scales

Cycloid (circular) scales have a smooth texture and are uniform, with a smooth outer edge or margin. They are most common on fish with soft fin rays, such as salmon and carp.

Cycloid (Circular) Scales

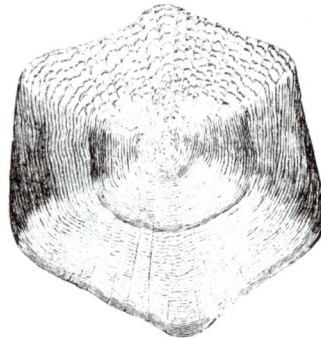

The cycloid scale of a carp has a smooth outer edge

This *Poropuntius huguenini* is a carp-like fish with circular cycloid scales that are smooth to the touch

Cycloid Scales

Cycloid (circular) scales have a smooth texture and are uniform, with a smooth outer edge or margin. They are most common on fish with soft fin rays, such as salmon and carp.

Cycloid (circular) scales are usually found on carp-like or salmon-like fishes			
bream	loach	minnow	grayling
bleak	chub	gudgeon	pike

Ctenoid Scales

Ctenoid (Toothed) Scales

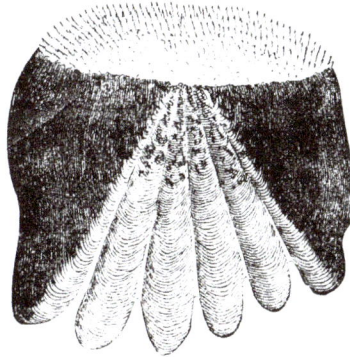

The ctenoid scale of a perch has a toothed outer edge (at top of image)

This dottyback is a perch-like fish with toothed ctenoid scales that are rough to the touch

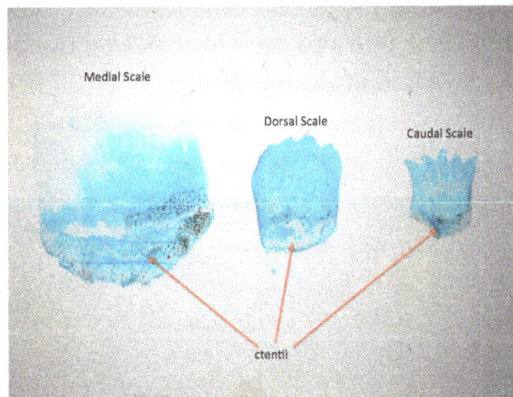

Three ctenoid scales from various locations of a perch were stained. Significant variation can be observed between the medial (middle of the fish), dorsal (top), and caudal (tail end) scales. The ctenii of each of the scales is labeled.

Ctenoid (toothed) scales are like cycloid scales, with small teeth along their outer edges. They are usually found on fishes with spiny fin rays, such as the perch-like fishes. The scales have a rough texture with a toothed outer or posterior edge featuring tiny teeth called ctenii. These scales contain almost no bone, being composed of a surface layer containing hydroxyapatite and calcium carbonate and a deeper layer composed mostly of collagen. The enamel of the other scale types is reduced to superficial ridges and ctenii.

Ctenoid (toothed) scales are usually found on perch-like fishes			
goby	flathead	scat	emperor

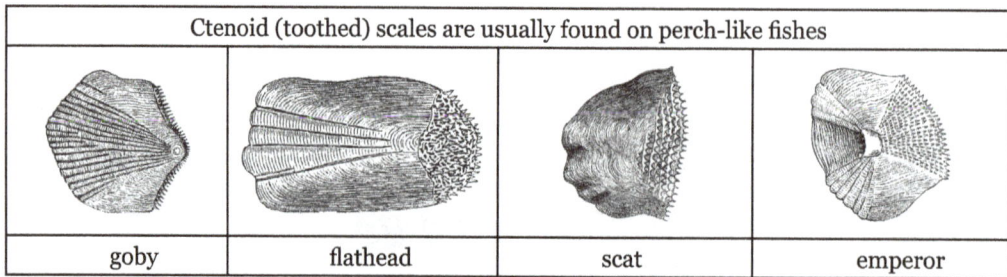

Ctenoid scales, similar to other epidermal structures, originate from placodes and distinctive cellular differentiation makes them exclusive from other structures that arise from the integument. Development starts near the caudal fin, along the lateral line of the fish. The development process begins with an accumulation of fibroblasts between the epidermis and dermis. Collagen fibrils begin to organize themselves in the dermal layer, which leads to the initiation of mineralization. The circumference of the scales grows first, followed by thickness when overlapping layers mineralize together.

Ctenoid scales can be further subdivided into three types:

- Crenate scales, where the margin of the scale bears indentations and projections.

- Spinoid scales, where the scale bears spines that are continuous with the scale itself.

- True ctenoid scales, where the spines on the scale are distinct structures.

Both cycloid and ctenoid scales are overlapping, making them more flexible than cosmoid and ganoid scales. Unlike ganoid scales, they grow in size through additions to the margin. The scales of some species exhibit bands of uneven seasonal growth called annuli (singular annulus). These bands can be used to age the fish. Most ray-finned fishes have ctenoid scales. Some species of flatfishes have ctenoid scales on the eyed side and cycloid scales on the blind side, while other species have ctenoid scales in males and cycloid scales in females.

Ganoid Scales

The longnose gar has diamond-shape ganoid scales

Ganoid scales are found in the sturgeons, paddlefishes, gars, bowfin, and bichirs. They are derived from cosmoid scales, with a layer of dentine in the place of cosmine, and a layer of inorganic bone salt called ganoine in place of vitrodentine. Most are diamond-shaped and connected by peg-and-socket joints. They are usually thick and have a minimal amount of overlap as compared to other scales. In sturgeons, the scales are greatly enlarged into armour plates along the sides and back, while in the bowfin the scales are greatly reduced in thickness to resemble cycloid scales.

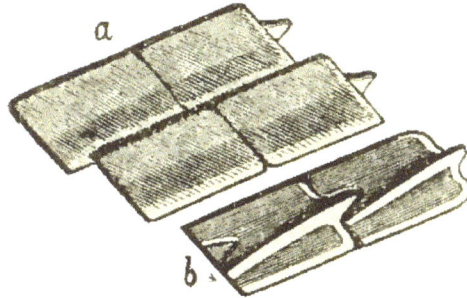

Ganoid scales of the Carboniferous fish, Amblypterus striatus. (a) shows the outer surface of four of the scales, and (b) shows the inner surface of two of the scales. Each of the rhomboidal ganoid scales of Amblypterus has a ridge on the inner surface which is produced at one end into a projecting peg which fits into a notch in the next scale, similar to the manner in which tiles are pegged together on the roof of a house.

Elasmoid Scales

Lobe-finned fishes, like this preserved coelacanth, have elasmoid scales

Elasmoid scales are thin, imbricated scales composed of a layer of dense, lamellar bone called isopedine, above which is a layer of tubercles usually composed of bone, as in *Eusthenopteron*. The layer of dentine that was present in the first sarcopterygians is usually reduced, as in the extant coelacanth, or entirely absent, as in extant lungfish and in the Devonian *Eusthenopteron*. Elasmoid scales have appeared several times over the course of fish evolution. They are present in some lobe-finned fishes: coelacanths, all extant and some extinct lungfishes, some tetrapodomorphs like *Eusthenopteron*, amiids, and teleosts, whose cycloid and ctenoid scales represent the least mineralized elasmoid scales.

Cosmoid Scales

Cosmoid scales are found in several ancient lobe-finned fishes, including some of the earliest lungfishes, and were probably derived from a fusion of placoid scales. They are composed of a layer of dense, lamellar bone called isopedine, above which is a layer of spongy bone supplied with blood vessels. The bone layers are covered by a complex dentine layer called cosmine and a superficial outer coating of vitrodentine. Cosmoid scales increase in size through the growth of the lamellar bone layer.

Scutes

A scute is another, less common, type of scale. Scute comes from Latin for *shield*, and can take the form of:

- an external shield-like bony plate, or

- a modified, thickened scale that often is keeled or spiny, or

- a projecting, modified (rough and strongly ridged) scale, usually associated with the lateral line, or on the caudal peduncle forming caudal keels, or along the ventral profile.

Some fish, such as pineconefish, are completely or partially covered in scutes. River herrings and threadfins have an abdominal row of scutes, which are scales with raised, sharp points that are used for protection. Some jacks have a row of scutes following the lateral line on either side.

Thelodont Scales

100µm

Left to right: denticles of *Paralogania* (?), *Shielia taiti*, *Lanarkia horrida*

The bony scales of thelodonts, the most abundant form of fossil fish, are well understood. The scales were formed and shed throughout the organisms' lifetimes, and quickly separated after their death.

Bone, a tissue that is both resistant to mechanical damage and relatively prone to fossilization, often preserves internal detail, which allows the histology and growth of the scales to be studied in detail. The scales comprise a non-growing "crown" composed of dentine, with a sometimes-ornamented enameloid upper surface and an aspidine base. Its growing base is made of cell-free bone, which sometimes developed anchorage structures to fix it in the side of the fish. Beyond that, there appear to be five types of bone-growth, which may represent five natural groupings within the thelodonts—or a spectrum ranging between the end members meta- (or ortho-) dentine and mesodentine tissues. Interestingly, each of the five scale morphs appears to resemble the scales of more derived groupings of fish, suggesting that thelodont groups may have been stem groups to succeeding clades of fish.

However, using scale morphology alone to distinguish species has some pitfalls. Within each organism, scale shape varies hugely according to body area, with intermediate forms appearing between different areas—and to make matters worse, scale morphology may not even be constant within one area. To confuse things further, scale morphologies are not unique to taxa, and may be indistinguishable on the same area of two different species.

The morphology and histology of thelodonts provides the main tool for quantifying their diversity and distinguishing between species, although ultimately using such convergent traits is prone to errors. Nonetheless, a framework comprising three groups has been proposed based upon scale morphology and histology.

Modifications

The cycloid scales of a common roach. The series of lateral line scales is visible in the lower half of the image.

Different groups of fish have evolved a number of modified scales to serve various functions.

- Almost all fishes have a lateral line, a system of mechanoreceptors that detect water movements. In bony fishes, the scales along the lateral line have central pores that allow water to contact the sensory cells.

- The dorsal fin spines of dogfish sharks and chimaeras, the stinging tail spines of stingrays, and the "saw" teeth of sawfishes and sawsharks are fused and modified placoid scales.

- Porcupine fishes have scales modified into spines.

- Surgeonfishes have a sharp, blade-like spines on either side of the caudal peduncle.

- Some herrings, anchovies, and halfbeaks have deciduous scales, which are easily shed and aid in escaping predators.

- Male Percina darters have a row of enlarged caducous scales between the pelvic fins and the anus.

Many groups of bony fishes, including pipefishes and seahorses, several families of catfishes, sticklebacks, and poachers, have developed external bony plates, structurally resembling placoid scales, as protective armour. In the boxfishes, the plates are all fused together to form a rigid shell enclosing the entire body. Yet these bony plates are not modified scales, but skin that has been ossified.

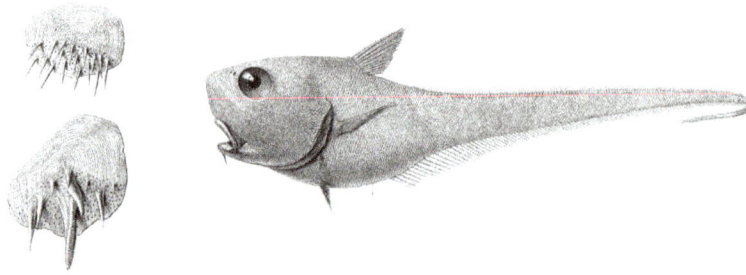

The size of the teeth on ctenoid scales can vary with position, as these scales from
the rattail Cetonurus crassiceps show

Eels seem scaleless, yet some species are covered with tiny smooth cycloid scales

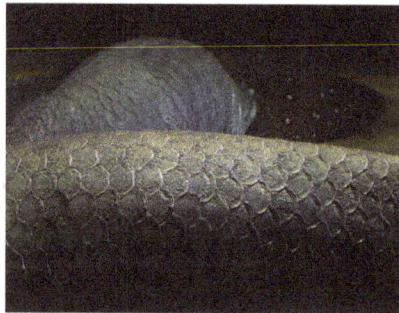

Scales of *Arapaima gigas*

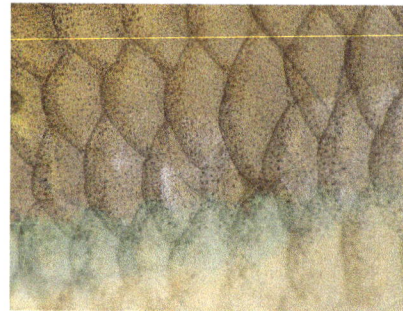

Scales of European bitterling from the hindflank

Fish Fin

Fins are usually the most distinctive features of a fish. They are composed of bony spines or rays protruding from the body with skin covering them and joining them together, either in a webbed fashion, as seen in most bony fish, or similar to a flipper, as seen in sharks. Apart from the tail or caudal fin, fish fins have no direct connection with the spine and are supported only by muscles. Their principal function is to help the fish swim. Fins located in different places on the fish serve different purposes such as moving forward, turning, keeping an upright position or stopping. Most fish use fins when swimming, flying fish use pectoral fins for gliding, and frogfish use them for crawling. Fins can also be used for other purposes; male sharks and mosquitofish use a modified fin to deliver sperm, thresher sharks use their caudal fin to stun prey, reef stonefish have spines in their dorsal fins that inject venom, anglerfish use the first spine of their dorsal fin like a fishing rod to lure prey, and triggerfish avoid predators by squeezing into coral crevices and using spines in their fins to lock themselves in place.

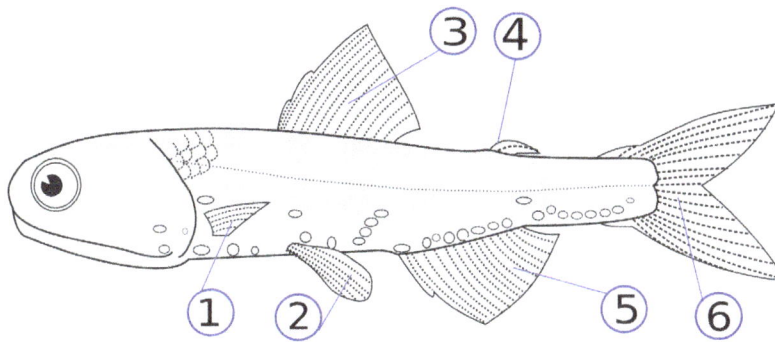

Ray fins on a teleost fish, Hector's lanternfish
(1) pectoral fins (paired), (2) pelvic fins (paired), (3) dorsal fin,
(4) adipose fin, (5) anal fin, (6) caudal (tail) fin

Types

For every type of fin, there are a number of fish species in which this particular fin has been lost during evolution.

Pectoral Fins

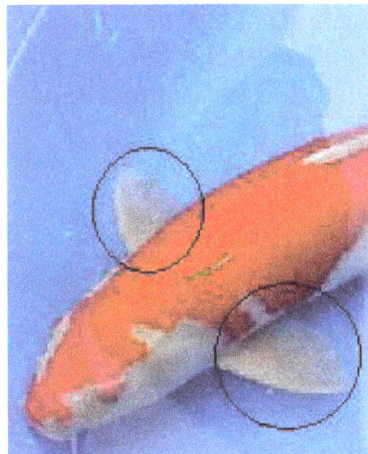

The paired pectoral fins are located on each side, usually just behind the operculum, and are homologous to the forelimbs of tetrapods.

- A peculiar function of pectoral fins, highly developed in some fish, is the creation of the dynamic lifting force that assists some fish, such as sharks, in maintaining depth and also enables the "flight" for flying fish.

- In many fish, the pectoral fins aid in walking, especially in the lobe-like fins of some anglerfish and in the mudskipper.

- Certain rays of the pectoral fins may be adapted into finger-like projections, such as in sea robins and flying gurnards.

    o   The "horns" of manta rays and their relatives are called cephalic fins; this is actually a modification of the anterior portion of the pectoral fin.

Pelvic Fins (Ventral Fins)

The paired pelvic or ventral fins are typically located ventrally below and behind the pectoral fins, although in many fish families they may be positioned in front of the pectoral fins (e.g. cods). They are homologous to the hindlimbs of tetrapods. The pelvic fin assists the fish in going up or down through the water, turning sharply, and stopping quickly.

- In gobies, the pelvic fins are often fused into a single sucker disk. This can be used to attach to objects.

- Pelvic fins can take many positions along the ventral surface of the fish. The ancestral *abdominal* position is seen in (for example) the minnows; the *thoracic* position in sunfish; and the *jugular* position, when the pelvics are anterior to the pectoral fins, as seen in the burbot.

Dorsal Fin

Dorsal fin of a shark

Dorsal fins are located on the back. A fish can have up to three dorsal fins. The dorsal fins serve to protect the fish against rolling, and assist it in sudden turns and stops.

- In anglerfish, the anterior of the dorsal fin is modified into an illicium and esca, a biological equivalent to a fishing rod and lure.

- The bones that support the dorsal fin are called *Pterygiophore*. There are two to three of them: "proximal", "middle", and "distal". In rock-hard, spinous fins the distal is often fused to the middle, or not present at all.

Dorsal fin of a chub (*Leuciscus cephalus*)

Anal/cloacal Fin

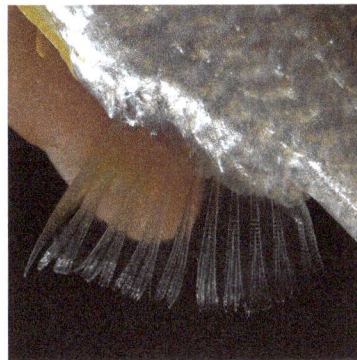

The anal/cloacal fin is located on the ventral surface behind the anus/cloaca. This fin is used to stabilize the fish while swimming.

Adipose Fin

Adipose fin of a trout

The adipose fin is a soft, fleshy fin found on the back behind the dorsal fin and just forward of the caudal fin. It is absent in many fish families, but found in nine of the 31 euteleostean orders (Percopsiformes, Myctophiformes, Aulopiformes, Stomiiformes, Salmoniformes, Osmeriformes, Characiformes, Siluriformes and Argentiniformes). Famous representatives of these orders are salmon, characids and catfish.

The function of the adipose fin is something of a mystery. It is frequently clipped off to mark hatchery-raised fish, though data from 2005 showed that trout with their adipose fin removed have an 8% higher tailbeat frequency. Additional information released in 2011 has suggested that the fin may be vital for the detection of, and response to, stimuli such as touch, sound and changes in pressure. Canadian researchers identified a neural network in the fin, indicating that it likely has a sensory function, but are still not sure exactly what the consequences of removing it are.

A comparative study in 2013 indicates the adipose fin can develop in two different ways. One is the salmoniform-type way, where the adipose fin develops from the larval-fin fold at the same time and in the same direct manner as the other median fins. The other is the characiform-type way, where the adipose fin develops late after the larval-fin fold has diminished and the other median fins have developed. They claim the existence of the characiform-type of development suggests the adipose fin is not "just a larval fin fold remainder" and is inconsistent with the view that the adipose fin lacks function.

Research published in 2014 indicates that the adipose fin has evolved repeatedly in separate lineages.

Caudal Fin (Tail Fin)

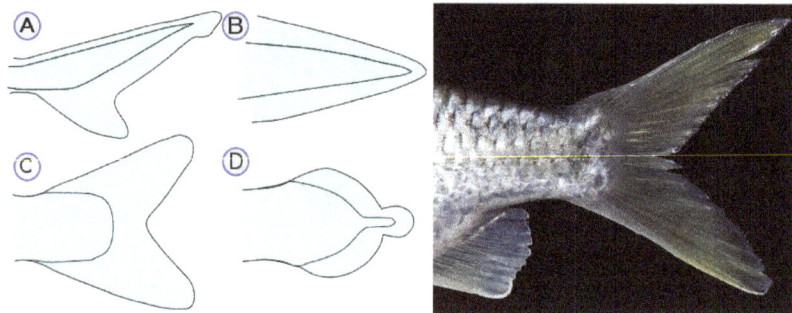

The caudal fin is the tail fin (from the Latin cauda meaning tail), located at the end of the caudal peduncle and is used for propulsion. See body-caudal fin locomotion.

(A) - Heterocercal means the vertebrae extend into the upper lobe of the tail, making it longer (as in sharks).

- Reversed heterocercal means that the vertebrae extend into the lower lobe of the tail, making it longer (as in the Anaspida)

(B) - Protocercal means the vertebrae extend to the tip of the tail and the tail is symmetrical but not expanded (as in amphioxus)

(C) - Homocercal where the fin appears superficially symmetric but in fact the vertebrae extend for a very short distance into the upper lobe of the fin

(D) - Diphycercal means the vertebrae extend to the tip of the tail and the tail is symmetrical and expanded (as in the bichir, lungfish, lamprey and coelacanth). Most Palaeozoic fishes had a diphycercal heterocercal tail.

Most modern fishes have a homocercal tail. These appear in a variety of shapes, and can appear:

- rounded

- truncated, ending in a more-or-less vertical edge (such as salmon)

- forked, ending in two prongs

- emarginate, ending with a slight inward curve.

- lunate or shaped like a crescent moon

Caudal Keel Finlets

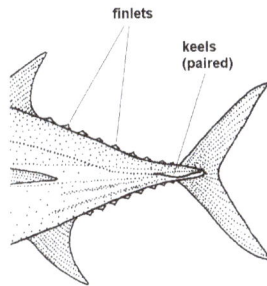

Drawing by Dr Tony Ayling

Some types of fast-swimming fish have a horizontal caudal keel just forward of the tail fin. Much like the keel of a ship, this is a lateral ridge on the caudal peduncle, usually composed of scutes (see below), that provides stability and support to the caudal fin. There may be a single paired keel, one on each side, or two pairs above and below.

Finlets are small fins, generally behind the dorsal and anal fins (in bichirs, there are only finlets on the dorsal surface and no dorsal fin). In some fish such as tuna or sauries, they are rayless, non-retractable, and found between the last dorsal and/or anal fin and the caudal fin.

Bony Fishes

Skeleton of a ray-finned fish

Bony fishes form a taxonomic group called Osteichthyes. They have skeletons made of bone, and can be contrasted with cartilaginous fishes which have skeletons made of cartilage. Bony fishes are divided into ray-finned and lobe-finned fish. Most fish are ray-finned, an extremely diverse and abundant group consisting of over 30,000 species. It is the largest class of vertebrates in existence today. In the distant past, lobe-finned fish were abundant. Nowadays they are mainly extinct, with only eight living species. Bony fish have fin spines and rays called lepidotrichia. They typically have swim bladders, which allows the fish to create a neutral balance between sinking and floating without having to use its fins. However, these are absent in many species, and have developed into primitive lungs in the lungfishes. Bony fishes also have an operculum, which helps them breathe without having to use fins to swim.

Lobe-finned

Lobe-finned fishes, like this coelacanth, have fins that are borne on a fleshy, lobelike, scaly stalk extending from the body. Due to the high number of fins it possesses, the coelacanth has high maneuverability and can orient their bodies in almost any direction in the water.

Lobe-finned fishes are a class of bony fishes called Sarcopterygii. They have fleshy, lobed, paired fins, which are joined to the body by a single bone. The fins of lobe-finned fish differ from those of all other fish in that each is borne on a fleshy, lobelike, scaly stalk extending from the body. Pectoral and pelvic fins have articulations resembling those of tetrapod limbs. These fins evolved into legs of the first tetrapod land vertebrates, amphibians. They also possess two dorsal fins with separate bases, as opposed to the single dorsal fin of ray-finned fish.

The coelacanth is another lobe-finned fish which is still extant. It is thought to have evolved into roughly its current form about 408 million years ago, during the early Devonian. Locomotion of the coelacanths is unique to their kind. To move around, coelacanths most commonly take advantage of up or downwellings of the current and drift. They use their paired fins to stabilize their movement through the water. While on the ocean floor their paired fins are not used for any kind of movement. Coelacanths can create thrust for quick starts by using their caudal fins. Due to the high number of fins they possess, coelacanths have high maneuverability and can orient their bodies in almost any direction in the water. They have been seen doing headstands and swimming belly up. It is thought that their rostral organ helps give the coelacanth electroperception, which aids in their movement around obstacles.

Ray-finned

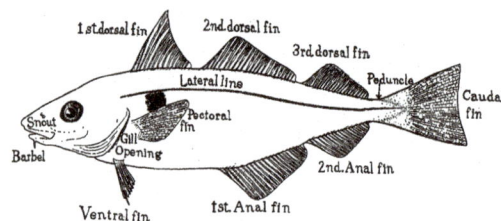

The haddock, a type of cod, is ray-finned. It has three dorsal and two anal fins

Ray-finned fishes are a class of bony fishes called Actinopterygii. Their fins contain spines or rays. A fin may contain only spiny rays, only soft rays, or a combination of both. If both are present, the spiny rays are always anterior. Spines are generally stiff and sharp. Rays are generally soft, flexible, segmented, and may be branched. This segmentation of rays is the main difference that separates them from spines; spines may be flexible in certain species, but they will never be segmented.

Spines have a variety of uses. In catfish, they are used as a form of defense; many catfish have the ability to lock their spines outwards. Triggerfish also use spines to lock themselves in crevices to prevent them being pulled out.

Lepidotrichia are bony, bilaterally paired, segmented fin rays found in bony fishes. They develop around actinotrichia as part of the dermal exoskeleton. Lepidotrichia are usually composed of bone, but in early osteichthyans such as *Cheirolepis*, there was also dentine and enamel. They are segmented and appear as a series of disks stacked one on top of another. The genetic basis for the formation of the fin rays is thought to be genes coded for the production of certain proteins. It has been suggested that the evolution of the tetrapod limb from lobe-finned fishes is related to the loss of these proteins.

Cartilaginous Fishes

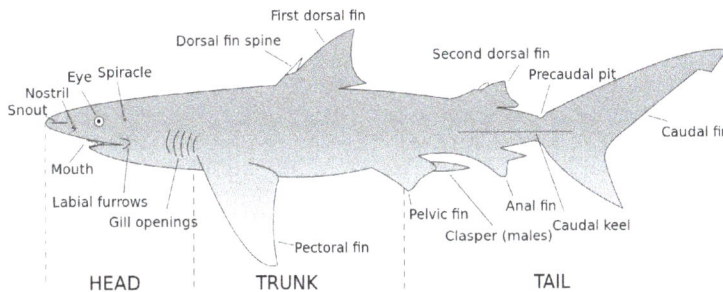

Cartilaginous fishes, like this shark, have fins that are elongated and supported with soft and unsegmented rays named ceratotrichia, filaments of elastic protein resembling the horny keratin in hair and feathers

Cartilaginous fishes are a class of fishes called Chondrichthyes. They have skeletons made of cartilage rather than bone. The class includes sharks, rays and chimaeras. Shark fin skeletons are elongated and supported with soft and unsegmented rays named ceratotrichia, filaments of elastic protein resembling the horny keratin in hair and feathers. Originally the pectoral and pelvic girdles, which do not contain any dermal elements, did not connect. In later forms, each pair of fins became ventrally connected in the middle when scapulocoracoid and pubioischiadic bars evolved. In rays, the pectoral fins have connected to the head and are very flexible. One of the primary characteristics present in most sharks is the heterocercal tail, which aids in locomotion. Most sharks have eight fins. Sharks can only drift away from objects directly in front of them because their fins do not allow them to move in the tail-first direction.

Caudal fin of a grey reef shark

As with most fish, the tails of sharks provide thrust, making speed and acceleration dependent on tail shape. Caudal fin shapes vary considerably between shark species, due to their evolution in separate environments. Sharks possess a heterocercal caudal fin in which the dorsal portion is usually noticeably larger than the ventral portion. This is because the shark's vertebral column extends into that dorsal portion, providing a greater surface area for muscle attachment. This allows more efficient locomotion among these negatively buoyant cartilaginous fish. By contrast, most bony fish possess a homocercal caudal fin.

Tiger sharks have a large upper lobe, which allows for slow cruising and sudden bursts of speed. The tiger shark must be able to twist and turn in the water easily when hunting to support its varied diet, whereas the porbeagle shark, which hunts schooling fish such as mackerel and herring, has a large lower lobe to help it keep pace with its fast-swimming prey. Other tail adaptations help sharks catch prey more directly, such as the thresher shark's usage of its powerful, elongated upper lobe to stun fish and squid.

Generating Thrust

Foil shaped fins generate thrust when moved, the lift of the fin sets water or air in motion and pushes the fin in the opposite direction. Aquatic animals get significant thrust by moving fins back and forth in water. Often the tail fin is used, but some aquatic animals generate thrust from pectoral fins.

Moving Fins can Provide Thrust

Fish get thrust moving vertical tail fins from side to side

Stingrays get thrust from large pectoral fins

Cavitation occurs when negative pressure causes bubbles (cavities) to form in a liquid, which then promptly and violently collapse. It can cause significant damage and wear. Cavitation damage can occur to the tail fins of powerful swimming marine animals, such as dolphins and tuna. Cavitation is more likely to occur near the surface of the ocean, where the ambient water pressure is relatively low. Even if they have the power to swim faster, dolphins may have to restrict their speed because collapsing cavitation bubbles on their tail are too painful. Cavitation also slows tuna, but for a different reason. Unlike dolphins, these fish do not feel the bubbles, because they have bony fins without nerve endings. Nevertheless, they cannot swim faster because the cavitation bubbles create a vapor film around their fins that limits their speed. Lesions have been found on tuna that are consistent with cavitation damage.

Scombrid fishes (tuna, mackerel and bonito) are particularly high-performance swimmers. Along the margin at the rear of their bodies is a line of small rayless, non-retractable fins, known as finlets. There has been much speculation about the function of these finlets. Research done in 2000 and 2001 by Nauen and Lauder indicated that "the finlets have a hydrodynamic effect on local flow during steady swimming" and that "the most posterior finlet is oriented to redirect flow into the developing tail vortex, which may increase thrust produced by the tail of swimming mackerel".

Fish use multiple fins, so it is possible that a given fin can have a hydrodynamic interaction with another fin. In particular, the fins immediately upstream of the caudal (tail) fin may be proximate fins that can directly affect the flow dynamics at the caudal fin. In 2011, researchers using volumetric imaging techniques were able to generate "the first instantaneous three-dimensional views of wake structures as they are produced by freely swimming fishes". They found that "continuous tail beats resulted in the formation of a linked chain of vortex rings" and that "the dorsal and anal fin wakes are rapidly entrained by the caudal fin wake, approximately within the timeframe of a subsequent tail beat".

Controlling Motion

Once motion has been established, the motion itself can be controlled with the use of other fins.

Specialised Fins are used to Control Motion

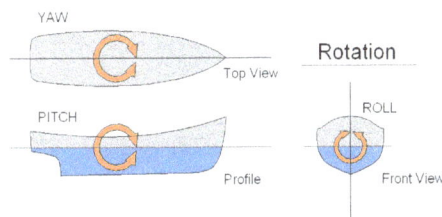

Like boats and airplanes, fish need some control over six degrees of freedom, three translational (heaving, swaying and surging) and three rotational (pitching, yawing and rolling)

Many reef fish have pectoral and pelvic fins optimised for flattened bodies

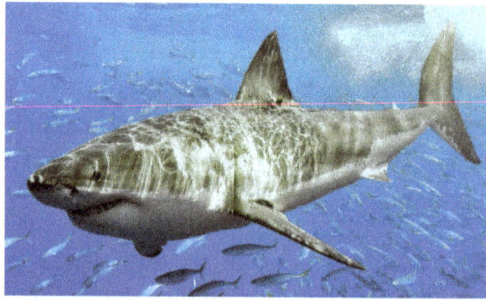

The dorsal fin of a white shark contain dermal fibers that work "like riggings that stabilize a ship's mast", and stiffen dynamically as the shark swims faster to control roll and yaw.

The bodies of reef fishes are often shaped differently from open water fishes. Open water fishes are usually built for speed, streamlined like torpedoes to minimise friction as they move through the water. Reef fish operate in the relatively confined spaces and complex underwater landscapes of coral reefs. For this manoeuvrability is more important than straight line speed, so coral reef fish have developed bodies which optimize their ability to dart and change direction. They outwit predators by dodging into fissures in the reef or playing hide and seek around coral heads. The pectoral and pelvic fins of many reef fish, such as butterflyfish, damselfish and angelfish, have evolved so they can act as brakes and allow complex manoeuvres. Many reef fish, such as butterflyfish, damselfish and angelfish, have evolved bodies which are deep and laterally compressed like a pancake, and will fit into fissures in rocks. Their pelvic and pectoral fins are designed differently, so they act together with the flattened body to optimise manoeuvrability. Some fishes, such as puffer fish, filefish and trunkfish, rely on pectoral fins for swimming and hardly use tail fins at all.

Reproduction

This male mosquitofish has a gonopodium, an anal fin which functions as an intromittent organ

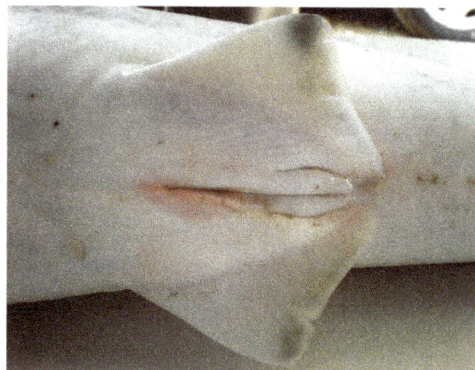

This young male spinner shark has claspers, a modification to the pelvic fins which also function as intromittent organs

Male cartilaginous fishes (sharks and rays), as well as the males of some live-bearing ray finned fishes, have fins that have been modified to function as intromittent organs, reproductive append-ages which allow internal fertilization. In ray finned fish they are called *gonopodia* or *andropodia*, and in cartilaginous fish they are called *claspers*.

Gonopodia are found on the males of some species in the Anablepidae and Poeciliidae families. They are anal fins that have been modified to function as movable intromittent organs and are used to impregnate females with milt during mating. The third, fourth and fifth rays of the male's anal fin are formed into a tube-like structure in which the sperm of the fish is ejected. When ready for mating, the gonopodium becomes erect and points forward towards the female. The male shortly inserts the organ into the sex opening of the female, with hook-like adaptations that allow the fish to grip onto the female to ensure impregnation. If a female remains stationary and her partner contacts her vent with his gonopodium, she is fertilized. The sperm is preserved in the female's oviduct. This allows females to fertilize themselves at any time without further assistance from males. In some species, the gonopodium may be half the total body length. Occasionally the fin is too long to be used, as in the "lyretail" breeds of *Xiphophorus helleri*. Hormone treated females may develop gonopodia. These are useless for breeding.

Similar organs with similar characteristics are found in other fishes, for example the *andropodium* in the *Hemirhamphodon* or in the Goodeidae.

Claspers are found on the males of cartilaginous fishes. They are the posterior part of the pelvic fins that have also been modified to function as intromittent organs, and are used to channel se-men into the female's cloaca during copulation. The act of mating in sharks usually includes rais-ing one of the claspers to allow water into a siphon through a specific orifice. The clasper is then inserted into the cloaca, where it opens like an umbrella to anchor its position. The siphon then begins to contract expelling water and sperm.

Other Uses

The Indo-Pacific sailfish has a prominent dorsal fin. Like scombroids and other billfish, they stream-line themselves by retracting their dorsal fins into a grove in their body when they swim. The huge dorsal fin, or sail, of the sailfish is kept retracted most of the time. Sailfish raise them if they want to herd a school of small fish, and also after periods of high activity, presumably to cool down.

Frogfish use their pectoral and pelvic fins to walk along the ocean bottom

The oriental flying gurnard has large pectoral fins which it normally holds against its body, and expands when threatened to scare predators. Despite its name, it is a demersal fish, not a flying fish, and uses its pelvic fins to walk along the bottom of the ocean.

Flying fish achieve sufficient lift to glide above the surface of the water thanks to their enlarged pectoral fins

Large retractable dorsal fin of the Indo-Pacific sailfish

Fins can have an adaptive significance as sexual ornaments. During courtship, the female cichlid, *Pelvicachromis taeniatus*, displays a large and visually arresting purple pelvic fin. "The researchers found that males clearly preferred females with a larger pelvic fin and that pelvic fins grew in a more disproportionate way than other fins on female fish."

The thresher shark uses its caudal fin to stun prey

Other uses of Fins

The Oriental flying gurnard has large pectoral fins with eye spots which it displays to scare predators

During courtship, the female cichlid, *Pelvicachromis taeniatus*, displays her visually arresting purple pelvic fin

Triggerfish squeeze into coral crevices to avoid predators, and lock themselves in place with the first spine of their dorsal fin

The first spine of the dorsal fin of the anglerfish is modified so it functions like a fishing rod with a lure

In some Asian countries shark fins are a culinary delicacy

Evolution

Evolution of Paired Fins

There are two prevailing hypotheses that have been historically debated as models for the evolution of paired fins in fish: the gill arch theory and the lateral fin-fold theory. The former, commonly referred to as the "Gegenbaur hypothesis," was posited in 1870 and proposes that the "paired fins are derived from gill structures". This fell out of popularity in favor of the lateral fin-fold theory, first suggested in 1877, which proposes that paired fins budded from longitudinal, lateral folds along the epidermis just behind the gills. There is weak support for both hypotheses in the fossil record and in embryology. However, recent insights from developmental patterning have prompted reconsideration of both theories in order to better elucidate the origins of paired fins.

Classical Theories

Karl Gegenbaur's concept of the "Archipterygium" was introduced in 1876. It was described as a gill ray, or "joined cartilaginous stem," that extended from the gill arch. Additional rays arose from along the arch and from the central gill ray. Gegenbaur suggested a model of transformative homology – that all vertebrate paired fins and limbs were transformations of the Archipterygium. Based on this theory, paired appendages such as pectoral and pelvic fins would have differentiated from the branchial arches and migrated posteriorly. However, there has been limited support for this hypothesis in the fossil record both morphologically and phylogenically. In addition, there was little to no evidence of an anterior-posterior migration of pelvic fins. Such shortcomings of the gill-arch theory led to its early demise in favor of the lateral fin-fold theory proposed by St. George Jackson Mivart, Francis Balfour, and James Kingsley Thacher.

The lateral fin-fold theory hypothesized that paired fins developed from lateral folds along the body wall of the fish. Just as segmentation and budding of the median fin fold gave rise to the median fins, a similar mechanism of fin bud segmentation and elongation from a lateral fin fold was proposed to have given rise to the paired pectoral and pelvic fins. However, there was little evidence of a lateral fold-to-fin transition in the fossil record. In addition, it was later demonstrated phylogenically that pectoral and pelvic fins arise from distinct evolutionary and mechanistic origins.

Evolutionary Developmental Biology

Recent studies in the ontogeny and evolution of paired appendages have compared finless vertebrates – such as lampreys – with chondricthyes, the most basal living vertebrate with paired fins. In 2006, researchers found that the same genetic programming involved in the segmentation and development of median fins was found in the development of paired appendages in catsharks. Although these findings do not directly support the lateral fin-fold hypothesis, the original concept of a shared median-paired fin evolutionary developmental mechanism remains relevant.

We find a similar renovation of an old theory in the developmental programming of chondrichthyan gill arches and paired appendages. In 2009, researchers at the University of Chicago demonstrated that there are shared molecular patterning mechanisms in the early development of the chondrichthyan gill arch and paired fins. Findings such as these have prompted reconsideration of the once-debunked gill-arch theory.

From Fins to Limbs

Fish are the ancestors of all mammals, reptiles, birds and amphibians. In particular, terrestrial tetrapods (four-legged animals) evolved from fish and made their first forays onto land 400 million years ago. They used paired pectoral and pelvic fins for locomotion. The pectoral fins developed into forelegs (arms in the case of humans) and the pelvic fins developed into hind legs. Much of the genetic machinery that builds a walking limb in a tetrapod is already present in the swimming fin of a fish.

Aristotle recognised the distinction between analogous and homologous structures, and made the following prophetic comparison: *"Birds in a way resemble fishes. For birds have their wings in the upper part of their bodies and fishes have two fins in the front part of their bodies. Birds have feet on their underpart and most fishes have a second pair of fins in their under-part and near their front fins."*

– Aristotle, De incessu animalium

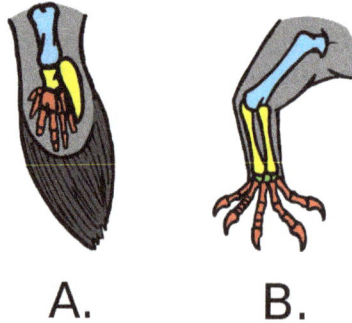

A.   B.

Comparison between A) the swimming fin of a lobe-finned fish and B) the walking leg of a tetrapod. Bones considered to correspond with each other have the same color.

In a parallel but independent evolution, the ancient reptile *Ichthyosaurus communis* developed fins (or flippers) very similar to fish (or dolphins)

In 2011, researchers at Monash University in Australia used primitive but still living lungfish "to trace the evolution of pelvic fin muscles to find out how the load-bearing hind limbs of the tetrapods evolved." Further research at the University of Chicago found bottom-walking lungfishes had already evolved characteristics of the walking gaits of terrestrial tetrapods.

In a classic example of convergent evolution, the pectoral limbs of pterosaurs, birds and bats further evolved along independent paths into flying wings. Even with flying wings there are many similarities with walking legs, and core aspects of the genetic blueprint of the pectoral fin have been retained.

The first mammals appeared during the Permian period (between 298.9 and 252.17 million years ago). Several groups of these mammals started returning to the sea, including the cetaceans (whales, dolphins and porpoises). Recent DNA analysis suggests that cetaceans evolved from within the even-toed ungulates, and that they share a common ancestor with the hippopotamus. About 23 million years ago another group of bearlike land mammals started returning to the sea. These were the seals. What had become walking limbs in cetaceans and seals evolved independently into new forms of swimming fins. The forelimbs became flippers, while the hindlimbs were either lost (cetaceans) or also modified into flipper (pinnipeds). In cetaceans, the tail gained two fins at the end, called a fluke. Fish tails are usually vertical and move from side to side. Cetacean flukes are horizontal and move up and down, because cetacean spines bend the same way as in other mammals.

Similar adaptations for fully aquatic lifestyle are found both in dolphins and ichthyosaurs

Ichthyosaurs are ancient reptiles that resembled dolphins. They first appeared about 245 million years ago and disappeared about 90 million years ago.

"This sea-going reptile with terrestrial ancestors converged so strongly on fishes that it actually evolved a dorsal fin and tail fin for improved aquatic locomotion. These structures are all the more remarkable because they evolved from nothing — the ancestral terrestrial reptile had no hump on its back or blade on its tail to serve as a precursor."

The biologist Stephen Jay Gould said the ichthyosaur was his favorite example of convergent evolution.

Fins or flippers of varying forms and at varying locations (limbs, body, tail) have also evolved in a number of other tetrapod groups, including diving birds such as penguins (modified from wings), sea turtles (forelimbs modified into flippers), mosasaurs (limbs modified into flippers), and sea snakes (vertically expanded, flattened tail fin).

Robotic Fins

In the 1990s, the CIA built a robotic catfish called *Charlie*, designed
to collect underwater intelligence undetected

The use of fins for the propulsion of aquatic animals can be remarkably effective. It has been calculated that some fish can achieve a propulsive efficiency greater than 90%. Fish can accelerate and maneuver much more effectively than boats or submarine, and produce less water disturbance and noise. This has led to biomimetic studies of underwater robots which attempt to emulate the locomotion of aquatic animals. An example is the Robot Tuna built by the Institute of Field Robotics, to analyze and mathematically model thunniform motion. In 2005, the Sea Life London Aquarium displayed three robotic fish created by the computer science department at the University of Essex. The fish were designed to be autonomous, swimming around and avoiding obstacles like real fish. Their creator claimed that he was trying to combine "the speed of tuna, acceleration of a pike, and the navigating skills of an eel."

The *AquaPenguin*, developed by Festo of Germany, copies the streamlined shape and propulsion by front flippers of penguins. Festo also developed *AquaRay*, *AquaJelly* and *AiraCuda*, respectively emulating the locomotion of manta rays, jellyfish and barracuda.

In 2004, Hugh Herr at MIT prototyped a biomechatronic robotic fish with a living actuator by surgically transplanting muscles from frog legs to the robot and then making the robot swim by pulsing the muscle fibers with electricity.

Robotic fish offer some research advantages, such as the ability to examine an individual part of a fish design in isolation from the rest of the fish. However, this risks oversimplifying the biology so key aspects of the animal design are overlooked. Robotic fish also allow researchers to vary a single parameter, such as flexibility or a specific motion control. Researchers can directly measure forces, which is not easy to do in live fish. "Robotic devices also facilitate three-dimensional kinematic studies and correlated hydrodynamic analyses, as the location of the locomotor surface can be known accurately. And, individual components of a natural motion (such as outstroke vs. instroke of a flapping appendage) can be programmed separately, which is certainly difficult to achieve when working with a live animal."

Fish Locomotion

Fish propel themselves through water using many different mechanisms.

Fish locomotion is the variety of types of animal locomotion used by fish, principally by swimming. This however is achieved in different groups of fish by a variety of mechanisms of propulsion in water, most often by wavelike movements of the fish's body and tail, and in various specialised fish by movements of the fins. The major forms of locomotion in fish are anguilliform, in which a wave passes evenly along a long slender body; sub-carangiform, in which the wave increases quickly in amplitude towards the tail; carangiform, in which the wave is concentrated near the tail, which oscillates rapidly; thunniform, rapid swimming with a large powerful crescent-shaped tail; and ostraciiform, with almost no oscillation except of the tail fin. More specialised fish include movement by pectoral fins with a mainly stiff body, as in the sunfish; and movement by propagating a wave along the long fins with a motionless body in fish with electric organs such as the knifefish.

In addition, some fish can variously "walk", i.e., move over land, burrow in mud, and glide through the air.

Swimming

Fish swim by exerting force against the surrounding water. There are exceptions, but this is normally achieved by the fish contracting muscles on either side of its body in order to generate waves of flexion that travel the length of the body from nose to tail, generally getting larger as they go along. The vector forces exerted on the water by such motion cancel out laterally, but generate a net force backwards which in turn pushes the fish forward through the water. Most fishes generate thrust using lateral movements of their body and caudal fin, but many other species move mainly using their median and paired fins. The latter group swim slowly, but can turn rapidly, as is needed when living in coral reefs for example. But they can't swim as fast as fish using their bodies and caudal fins.

Body/caudal Fin Propulsion

There are five groups that differ in the fraction of their body that is displaced laterally:

Anguilliform

Eels propagate a more or less constant-sized flexion wave along their slender bodies.

In the anguilliform group, containing some long, slender fish such as eels, there is little increase in the amplitude of the flexion wave as it passes along the body.

Sub-carangiform

The sub-carangiform group has a more marked increase in wave amplitude along the body with the vast majority of the work being done by the rear half of the fish. In general, the fish body is stiffer, making for higher speed but reduced maneuverability. Trout use sub-carangiform locomotion.

Carangiform

The carangiform group, named for the Carangidae, are stiffer and faster-moving than the previous groups. The vast majority of movement is concentrated in the very rear of the body and tail. Carangiform swimmers generally have rapidly oscillating tails.

Thunniform

Tunas such as the bluefin swim fast with their large crescent-shaped tails.

The thunniform group contains high-speed long-distance swimmers, and is a unique trait (an autapomorphy) of the tunas. Here, virtually all the sideways movement is in the tail and the region connecting the main body to the tail (the peduncle). The tail itself tends to be large and crescent shaped.

Ostraciiform

The ostraciiform group have no appreciable body wave when they employ caudal locomotion. Only the tail fin itself oscillates (often very rapidly) to create thrust. This group includes Ostraciidae.

Median/paired Fin Propulsion

Boxfish use median-paired fin swimming, as they are not well streamlined, and use primarily their pectoral fins to produce thrust.

Not all fish fit comfortably in the above groups. Ocean sunfish, for example, have a completely different system, the tetraodontiform mode, and many small fish use their pectoral fins for swimming as well as for steering and dynamic lift. Fish with electric organs, such as those in the knifefish (Gymnotiformes), swim by undulating their very long fins while keeping the body still, presumably so as not to disturb the electric field that they generate.

Many fish swim using combined behavior of their two pectoral fins or both their anal and dorsal fins. Different types of Median paired fin propulsion can be achieved by preferentially using one fin pair over the other, and include rajiform, diodontiform, amiiform, gymnotiform and balistiform modes.

Rajiform

Rajiform locomotion is characteristic of rays, skates, and mantas when thrust is produced by vertical undulations along large, well developed pectoral fins.

Diodontiform

Porcupine fish (here, *Diodon nicthemerus*) swim by undulating their pectoral fins.

Diodontiform locomotion propels the fish propagating undulations along large pectoral fins, as seen in the porcupinefish (Diodontidae).

Amiiform

Amiiform locomotion consists of undulations of a long dorsal fin while the body axis is held straight and stable, as seen in the bowfin.

Gymnotiform

Gymnotus maintains a straight back while swimming to avoid disturbing its electric sense.

Gymnotiform locomotion consists of undulations of a long anal fin, essentially upside down amiiform, seen in the knifefish (Gymnotiformes).

Balistiform

In balistiform locomotion, both anal and dorsal fins undulate, as seen in the Zeidae.

Oscillatory

Oscillation is viewed as pectoral-fin-based swimming and is best known as mobuliform locomotion. The motion can be described as the production of less than half a wave on the fin, similar to a bird wing flapping. Pelagic stingrays, such as the manta, cownose, eagle and bat rays use oscillatory locomotion.

Tetraodontiform

In tetraodontiform locomotion, the dorsal and anal fins are flapped as a unit, either in phase or exactly opposing one another, as seen in the Tetraodontiformes (boxfishes and pufferfishes). The ocean sunfish displays an extreme example of this mode.

Labriform

In labriform locomotion, seen in the wrasses (Labriformes), oscillatory movements of pectoral fins are either drag based or lift based. Propulsion is generated either as a reaction to drag produced by dragging the fins through the water in a rowing motion, or via lift mechanisms.

Dynamic Lift

Sharks are denser than water, and must swim continually, using dynamic lift from their pectoral fins.

Bone and muscle tissues of fish are denser than water. To maintain depth fish such as sharks, but also some bony fish, increase buoyancy by means of a gas bladder or by storing oils or lipids. Fish without these features use dynamic lift instead. It is done using their pectoral fins in a manner similar to the use of wings by airplanes and birds. As these fish swim, their pectoral fins are positioned to create lift which allows the fish to maintain a certain depth. The two major drawbacks of this method are that these fish must stay moving to stay afloat and that they are incapable of swimming backwards or hovering.

Hydrodynamics

Similarly to the aerodynamics of flight, powered swimming requires animals to overcome drag by producing thrust. Unlike flying, however, swimming animals often do not need to supply much vertical force because the effect of buoyancy can counter the downward pull of gravity, allowing these animals to float without much effort. While there is great diversity in fish locomotion, swimming behavior can be classified into two distinct "modes" based on the body structures involved in thrust production, Median-Paired Fin (MPF) and Body-Caudal Fin (BCF). Within each of these classifications, there are numerous specifications along a spectrum of behaviours from purely undulatory to entirely oscillatory. In undulatory swimming modes, thrust is produced by wave-like movements of the propulsive structure (usually a fin or the whole body). Oscillatory modes, on the other hand, are characterized by thrust produced by swiveling of the propulsive structure on an attachment point without any wave-like motion.

Body-caudal Fin

Most fish swim by generating undulatory waves that propagate down the body through the caudal fin. This form of undulatory locomotion is termed Body-Caudal Fin (BCF) swimming on the basis of the body structures used; it includes anguilliform, sub-carangiform, carangiform, and thunniform locomotory modes, as well as the oscillatory ostraciiform mode.

Adaptation

Similar to adaptation in avian flight, swimming behaviors in fish can be thought of as a balance of stability and maneuverability. Because BCF swimming relies on more caudal body structures that can direct powerful thrust only rearwards, this form of locomotion is particularly effective for accelerating quickly and cruising continuously. BCF swimming is, therefore, inherently stable and is often seen in fish with large migration patterns that must maximize efficiency over long periods. Propulsive forces in MPF swimming, on the other hand, are generated by multiple fins located on either side of the body that can be coordinated to execute elaborate turns. As a result, MPF swimming is well adapted for high maneuverability and is often seen in smaller fish that require elaborate escape patterns.

The habitats occupied by fishes are often related to their swimming capabilities. On coral reefs, the faster-swimming fish species typically live in wave-swept habitats subject to fast water flow speeds, while the slower fishes live in sheltered habitats with low levels of water movement.

Fish do not rely exclusively on one locomotor mode, but are rather locomotor generalists, choosing among and combining behaviors from many available behavioral techniques. At slower speeds, predominantly BCF swimmers often incorporate movement of their pectoral, anal, and dorsal fins as an additional stabilizing mechanism at slower speeds, but hold them close to their body at high speeds to improve streamlining and reducing drag. Zebrafish have even been observed to alter their locomotor behavior in response to changing hydrodynamic influences throughout growth and maturation.

In addition to adapting locomotor behavior, controlling buoyancy effects is critical for aquatic survival since aquatic ecosystems vary greatly by depth. Fish generally control their depth by regulating the amount of gas in specialized organs that are much like balloons. By changing the amount of gas in these swim bladders, fish actively control their density. If they increase the amount of air in their swim bladder, their overall density will become less than the surrounding water, and increased upward buoyancy pressures will cause the fish to rise until they reach a depth at which they are again at equilibrium with the surrounding water.

Flight

The transition of predominantly swimming locomotion directly to flight has evolved in a single family of marine fish, the Exocoetidae. Flying fish are not true fliers in the sense that they do not execute powered flight. Instead, these species glide directly over the surface of the ocean water without ever flapping their "wings." Flying fish have evolved abnormally large pectoral fins that act as airfoils and provide lift when the fish launches itself out of the water. Additional forward thrust and steering forces are created by dipping the hypocaudal (i.e. bottom) lobe of their caudal fin into the water and vibrating it very quickly, in contrast to diving birds in which these forces are produced by the same locomotor module used for propulsion. Of the 64 extant species of flying fish, only two distinct body plans exist, each of which optimizes two different behaviors.

Flying fish gain sufficient lift to glide above the water thanks to their enlarged pectoral fins.

Tradeoffs

While most fish have caudal fins with evenly sized lobes (i.e. homocaudal), flying fish have an enlarged ventral lobe (i.e. hypocaudal) which facilitates dipping only a portion of the tail back onto the water for additional thrust production and steering.

Because flying fish are primarily aquatic animals, their body density must be close to that of water for buoyancy stability. This primary requirement for swimming, however, means that flying fish are heavier (have a larger mass) than other habitual fliers, resulting in higher wing loading and lift to drag ratios for flying fish compared to a comparably sized bird. Differences in wing area, wing span, wing loading, and aspect ratio have been used to classify flying fish into two distinct classifications based on these different aerodynamic designs.

Biplane Body Plan

In the biplane or *Cypselurus* body plan, both the pectoral and pelvic fins are enlarged to provide lift during flight. These fish also tend to have "flatter" bodies which increase the total lift producing area thus allowing them to "hang" in the air better than more streamlined shapes. As a result of this high lift production, these fish are excellent gliders and are well adapted for maximizing flight distance and duration.

Comparatively, *Cypselurus* flying fish have lower wing loading and smaller aspect ratios (i.e. broader wings) than their *Exocoetus* monoplane counterparts, which contributes to their ability to fly for longer distances than fish with this alternative body plan. Flying fish with the biplane design take advantage of their high lift production abilities when launching from the water by utilizing a "taxiing glide" in which the hypocaudal lobe remains in the water to generate thrust even after the trunk clears the water's surface and the wings are opened with a small angle of attack for lift generation.

Monoplane Body Plan

In the monoplane body plan of *Exocoetus*, only the pectoral fins are abnormally large, while the pelvic fins are small.

In the *Exocoetus* or monoplane body plan, only the pectoral fins are enlarged to provide lift. Fish with this body plan tend to have a more streamlined body, higher aspect ratios (long, narrow wings), and higher wing loading than fish with the biplane body plan, making these fish well adapted for higher flying speeds. Flying fish with a monoplane body plan demonstrate different launching behaviors from their biplane counterparts. Instead of extending their duration of thrust production, monoplane fish launch from the water at high speeds at a large angle of attack (sometimes up to 45 degrees). In this way, monoplane fish are taking advantage of their adaptation for high flight speed, while fish with biplane designs exploit their lift production abilities during takeoff.

Walking

A "walking fish" is a fish that is able to travel over land for extended periods of time. Some other cases of nonstandard fish locomotion include fish "walking" along the sea floor, such as the handfish or frogfish.

Most commonly, walking fish are amphibious fish. Able to spend longer times out of water, these fish may use a number of means of locomotion, including springing, snake-like lateral undulation, and tripod-like walking. The mudskippers are probably the best land-adapted of contemporary fish and are able to spend days moving about out of water and can even climb mangroves, although to only modest heights. The Climbing gourami is often specifically referred to as a "walking fish", although it does not actually "walk", but rather moves in a jerky way by supporting itself on the extended edges of its gill plates and pushing itself by its fins and tail. Some reports indicate that it can also climb trees.

There are a number of fish that are less adept at actual walking, such as the walking catfish. Despite being known for "walking on land", this fish usually wriggles and may use its pectoral fins to aid in its movement. Walking Catfish have a respiratory system that allows them to live out of water for several days. Some are invasive species. A notorious case in the United States is the Northern snakehead. Polypterids have rudimentary lungs and can also move about on land, though rather clumsily. The Mangrove rivulus can survive for months out of water and can move to places like hollow logs.

Ogcocephalus parvus

There are some species of fish that can "walk" along the sea floor but not on land; one such animal is the flying gurnard (it does not actually fly, and should not be confused with flying fish). The batfishes of the Ogcocephalidae family are also capable of walking along the sea floor. Bathypterois grallator, also known as a "tripodfish", stands on its three fins on the bottom of the ocean and hunts for food. The African lungfish (*P. annectens*) can use its fins to *"walk"* along the bottom of its tank in a manner similar to the way amphibians and land vertebrates use their limbs on land.

Burrowing

Many fishes, particularly eel-shaped fishes such as true eels, moray eels, and spiny eels, are capable of burrowing through sand or mud. Ophichthids, the snake eels, are capable of burrowing either forwards or backwards.

Swim Bladder

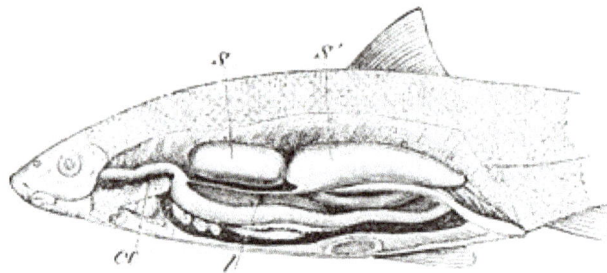

Internal positioning of the swim bladder of a bleak S: anterior, S': posterior portion of the air bladder œ: œsophagus; l: air passage of the air bladder

The swim bladder, gas bladder, fish maw or air bladder is an internal gas-filled organ that contributes to the ability of many bony fish (but not cartilaginous fish) to control their buoyancy, and thus to stay at their current water depth without having to waste energy in swimming. Also, the dorsal position of the swim bladder means the center of mass is below the center of volume, allowing it to act as a stabilizing agent. Additionally, the swim bladder functions as a resonating chamber, to produce or receive sound.

The swim bladder is evolutionarily homologous to the lungs. Charles Darwin remarked upon this in *On the Origin of Species*. Darwin reasoned that the lung in air-breathing vertebrates had derived from a more primitive swim bladder, but scientists now believe that the swim bladder derived from a more primitive lung.

In the embryonic stages some species, such as redlip blenny, have lost the swim bladder again, mostly bottom dwellers like the weather fish. Other fish like the Opah and the Pomfret use their pectoral fins to swim and balance the weight of the head to keep a horizontal position. The normally bottom dwelling sea robin can use their pectoral fins to produce lift while swimming.

The gas/tissue interface at the swim bladder produces a strong reflection of sound, which is used in sonar equipment to find fish.

The cartilaginous fish (e.g. sharks and rays) do not have swim bladders. Some of them can control their depth only by swimming (using dynamic lift); others store fats or oils with density less than that of seawater to produce a neutral or near neutral buoyancy, which does not change with depth.

Structure and Function

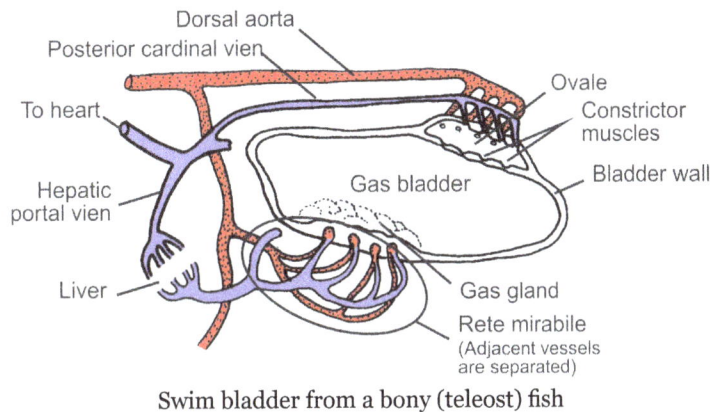

Swim bladder from a bony (teleost) fish

The swim bladder normally consists of two gas-filled sacs located in the dorsal portion of the fish, although in a few primitive species, there is only a single sac. It has flexible walls that contract or expand according to the ambient pressure. The walls of the bladder contain very few blood vessels and are lined with guanine crystals, which make them impermeable to gases. By adjusting the gas pressurising organ using the gas gland or oval window the fish can obtain neutral buoyancy and ascend and descend to a large range of depths. Due to the dorsal position it gives the fish lateral stability.

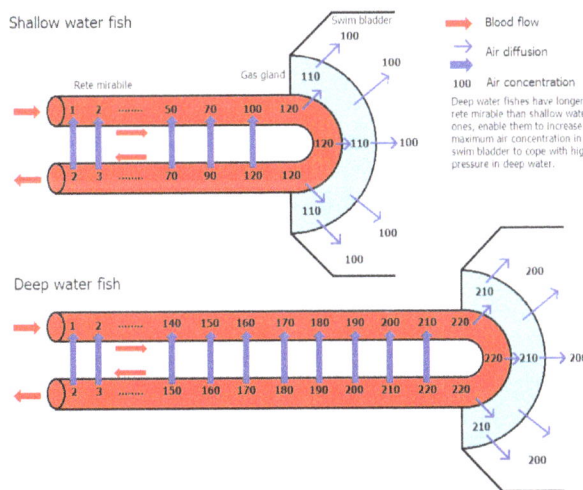

How gas is pumped into the swim bladder using counter-current exchange.

In physostomous swim bladders, a connection is retained between the swim bladder and the gut, the pneumatic duct, allowing the fish to fill up the swim bladder by "gulping" air. Excess gas can be removed in a similar manner.

In more derived varieties of fish (the physoclisti) the connection to the digestive tract is lost. In early life stages, these fish must rise to the surface to fill up their swim bladders; in later stages, the pneumatic duct disappears, and the gas gland has to introduce gas (usually oxygen) to the bladder to increase its volume and thus increase buoyancy. In order to introduce gas into the bladder, the gas gland excretes lactic acid and produces carbon dioxide. The resulting acidity causes the hemoglobin of the blood to lose its oxygen (Root effect) which then diffuses partly into the swim bladder. The blood flowing back to the body first enters a rete mirabile where virtually all the excess carbon dioxide and oxygen produced in the gas gland diffuses back to the arteries supplying the gas gland. Thus a very high gas pressure of oxygen can be obtained, which can even account for the presence of gas in the swim bladders of deep sea fish like the eel, requiring a pressure of hundreds of bars. Elsewhere, at a similar structure known as the *oval window*, the bladder is in contact with blood and the oxygen can diffuse back out again. Together with oxygen, other gases are salted out in the swim bladder which accounts for the high pressures of other gases as well.

The combination of gases in the bladder varies. In shallow water fish, the ratios closely approximate that of the atmosphere, while deep sea fish tend to have higher percentages of oxygen. For instance, the eel *Synaphobranchus* has been observed to have 75.1% oxygen, 20.5% nitrogen, 3.1% carbon dioxide, and 0.4% argon in its swim bladder.

Physoclist swim bladders have one important disadvantage: they prohibit fast rising, as the bladder would burst. Physostomes can "burp" out gas, though this complicates the process of re-submergence.

The swim bladder in some species, mainly fresh water fishes (Common carp, catfish, bowfin) is interconnected with the inner ear of the fish. They are connected by four bones called the Weberian ossicles from the Weberian apparatus. These bones can carry the vibrations to the Saccule and the Lagena (anatomy). They are suited for detecting sound and vibrations due to its low density in comparison to the density of the fish's body tissues. This increases the ability of sound detection. The swim bladder can radiate the pressure of sound which help increase its sensitivity and expand its hearing. In some deep sea fishes like the *Antimora*, the swim bladder maybe also connected to the Macula of saccule in order for the inner ear to receive a sensation from the sound pressure. In red-bellied piranha, the swimbladder may play an important role in sound production as a resonator. The sounds created by piranhas are generated through rapid contractions of the sonic muscles and is associated with the swimbladder.

Evolution

The West African lungfish possesses a lung homologous to swim bladders

The illustration of the swim bladder in fishes ... shows us clearly the highly important fact that an organ originally constructed for one purpose, namely, flotation, may be converted into one for a widely different purpose, namely, respiration. The swim bladder has, also, been worked in as an accessory to the auditory organs of certain fishes. All physiologists admit that the swimbladder is homologous, or "ideally similar" in position and structure with the lungs of the higher vertebrate animals: hence there is no reason to doubt that the swim bladder has actually been converted into lungs, or an organ used exclusively for respiration. According to this view it may be inferred that all vertebrate animals with true lungs are descended by ordinary generation from an ancient and unknown prototype, which was furnished with a floating apparatus or swim bladder.

Charles Darwin, 1859

Swim bladders are evolutionarily closely related (i.e., homologous) to lungs. Traditional wisdom has long held that the first lungs, simple sacs connected to the gut that allowed the organism to gulp air under oxygen-poor conditions, evolved into the lungs of today's terrestrial vertebrates and some fish (e.g., lungfish, gar, and bichir) and into the swim bladders of the ray-finned fish. In 1997, Farmer proposed that lungs evolved to supply the heart with oxygen. In fish, blood circulates from the gills to the skeletal muscle, and only then to the heart. During intense exercise, the oxygen in the blood gets used by the skeletal muscle before the blood reaches the heart. Primitive lungs gave an advantage by supplying the heart with oxygenated blood via the cardiac shunt. This theory is robustly supported by the fossil record, the ecology of extant air-breathing fishes, and the physiology of extant fishes. In embryonal development, both lung and swim bladder originate as an outpocketing from the gut; in the case of swim bladders, this connection to the gut continues to exist as the pneumatic duct in the more "primitive" ray-finned fish, and is lost in some of the more derived teleost orders. There are no animals which have both lungs and a swim bladder.

The cartilaginous fish (e.g., sharks and rays) split from the other fishes about 420 million years ago, and lack both lungs and swim bladders, suggesting that these structures evolved after that split. Correspondingly, these fish also have both heterocercal and pectoral fins which provide the necessary lift needed due to the lack of swim bladders. Teleost fish with swim bladders have neutral buoyancy, and have no need for this lift.

Deep Scattering Layer

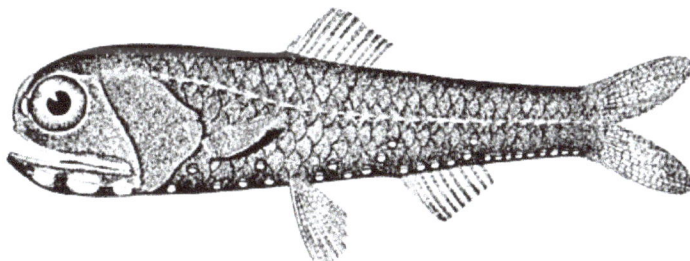

Most mesopelagic fishes are small filter feeders which ascend at night using their swimbladders to feed in the nutrient rich waters of the epipelagic zone. During the day, they return to the dark, cold, oxygen deficient waters of the mesopelagic where they are relatively safe from predators. Lanternfish account for as much as 65 percent of all deep sea fish biomass and are largely responsible for the deep scattering layer of the world's oceans.

Sonar operators, using the newly developed sonar technology during World War II, were puzzled by what appeared to be a false sea floor 300–500 metres deep at day, and less deep at night. This turned out to be due to millions of marine organisms, most particularly small mesopelagic fish, with swimbladders that reflected the sonar. These organisms migrate up into shallower water at dusk to feed on plankton. The layer is deeper when the moon is out, and can become shallower when clouds obscure the moon.

Most mesopelagic fish make daily vertical migrations, moving at night into the epipelagic zone, often following similar migrations of zooplankton, and returning to the depths for safety during the day. These vertical migrations often occur over large vertical distances, and are undertaken with the assistance of a swim bladder. The swim bladder is inflated when the fish wants to move up, and, given the high pressures in the mesoplegic zone, this requires significant energy. As the fish ascends, the pressure in the swimbladder must adjust to prevent it from bursting. When the fish wants to return to the depths, the swimbladder is deflated. Some mesopelagic fishes make daily migrations through the thermocline, where the temperature changes between 10 and 20°C, thus displaying considerable tolerance for temperature change.

Sampling via deep trawling indicates that lanternfish account for as much as 65% of all deep sea fish biomass. Indeed, lanternfish are among the most widely distributed, populous, and diverse of all vertebrates, playing an important ecological role as prey for larger organisms. The estimated global biomass of lanternfish is 550–660 million metric tonnes, several times the entire world fisheries catch. Lanternfish also account for much of the biomass responsible for the deep scattering layer of the world's oceans. Sonar reflects off the millions of lanternfish swim bladders, giving the appearance of a false bottom.

Human Uses

In some Asian cultures, the swim bladders of certain large fishes are considered a food delicacy. In China they are known as *fish maw* and are served in soups or stews.

The vanity price of a vanishing kind of maw is behind the imminent extinction of the vaquita, the world's smallest dolphin breed. Only found in Mexico's Gulf of California, the once numerous vaquita now number less than 60 in total. Vaquita die in gillnets set to catch totoaba (the world's largest drum fish). Totoaba are being hunted to extinction for its maw, which can sell for as much $10,000 per kilogram.

Swim bladders are also used in the food industry as a source of collagen. They can be made into a strong, water-resistant glue, or used to make isinglass for the clarification of beer. In earlier times they were used to make condoms.

Swim Bladder Disease

Swim bladder disease is a common ailment in aquarium fish. A fish with swim bladder disorder can float nose down tail up, or can float to the top or sink to the bottom of the aquarium.

Risk of Injury

Many anthropogenic activities, like pile driving or even Seismic wave, that could result from climate change or natural causes, can create high-intensity sound waves that cause a certain amount of damage to fishes that possess a gas bladder. Physostomes can release air in order to decrease the tension in the gas bladder that may cause internal injuries to other vital organs. While physoclisti can't expel air fast enough, making it more difficult to avoid any major injuries. Some of the commonly seen injuries included ruptured gas bladder and renal Haemorrhage. These mostly affect the overall health of the fish and didn't affect their mortality rate. Investigators used the High-Intensity-Controlled Impedance Fluid Filled (HICI-FT), a stainless-steel wave tube with a electromagnetic shaker. It simulates high-energy sound waves in aquatic far-field, plane-wave acoustic conditions.

Similar Structures in other Organisms

Siphonophores have a special swim bladder that allows the jellyfish-like colonies to float along the surface of the water while their tentacles trail below. This organ is unrelated to the one in fish.

Sensory Systems in Fish

Most fish possess highly developed sense organs. Nearly all daylight fish have color vision that is at least as good as a human's (see vision in fishes). Many fish also have chemoreceptors that are responsible for extraordinary senses of taste and smell. Although they have ears, many fish may not hear very well. Most fish have sensitive receptors that form the lateral line system, which detects gentle currents and vibrations, and senses the motion of nearby fish and prey. Sharks can sense frequencies in the range of 25 to 50 Hz through their lateral line.

Fish orient themselves using landmarks and may use mental maps based on multiple landmarks or symbols. Fish behavior in mazes reveals that they possess spatial memory and visual discrimination.

Vision

Fish have a refractive index gradient within the lens of their eyes which compensates for spherical aberration. Unlike humans, most fish adjust focus by moving the lens closer or further from the retina.

Vision is an important sensory system for most species of fish. Fish eyes are similar to those of terrestrial vertebrates like birds and mammals, but have a more spherical lens. Their retinas generally have both rod cells and cone cells (for scotopic and photopic vision), and most species have colour vision. Some fish can see ultraviolet and some can see polarized light. Amongst jawless fish, the lamprey has well-developed eyes, while the hagfish has only primitive eyespots. Fish vision shows adaptation to their visual environment, for example deep sea fishes have eyes suited to the dark environment.

Fish and other aquatic animals live in a different light environment than terrestrial species. Water absorbs light so that with increasing depth the amount of light available decreases quickly. The optic properties of water also lead to different wavelengths of light being absorbed to different degrees, for example light of long wavelengths (e.g. red, orange) is absorbed quite quickly compared to light of short wavelengths (blue, violet), though ultraviolet light (even shorter wavelength than blue) is absorbed quite quickly as well. Besides these universal qualities of water, different bodies of water may absorb light of different wavelengths because of salts and other chemicals in the water.

Hearing

Hearing is an important sensory system for most species of fish. Hearing threshold and the ability to localize sound sources are reduced underwater, in which the speed of sound is faster than in air. Underwater hearing is by bone conduction, and localization of sound appears to depend on differences in amplitude detected by bone conduction. Aquatic animals such as fish, however, have a more specialized hearing apparatus that is effective underwater.

Fish can sense sound through their lateral lines and their otoliths (ears). Some fishes, such as some species of carp and herring, hear through their swim bladders, which function rather like a hearing aid.

Hearing is well-developed in carp, which have the Weberian organ, three specialized vertebral processes that transfer vibrations in the swim bladder to the inner ear.

Although it is hard to test sharks' hearing, they may have a sharp sense of hearing and can possibly hear prey many miles away. A small opening on each side of their heads (not the spiracle) leads directly into the inner ear through a thin channel. The lateral line shows a similar arrangement, and is open to the environment via a series of openings called lateral line pores. This is a reminder of the common origin of these two vibration- and sound-detecting organs that are grouped together as the acoustico-lateralis system. In bony fish and tetrapods the external opening into the inner ear has been lost.

Current Detection

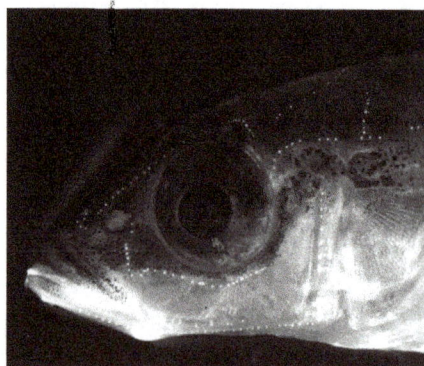

A *three-spined stickleback* with stained neuromasts

Hair cells in fish are used to detect water movements around their bodies. These hair cells are embedded in a jelly-like protrusion called cupula. The hair cells therefore can not be seen and do not appear on the surface of skin.

The lateral line in fish and aquatic forms of amphibians is a detection system of water currents, consisting mostly of vortices. The lateral line is also sensitive to low-frequency vibrations. The mechanoreceptors are hair cells, the same mechanoreceptors for vestibular sense and hearing. It is used primarily for navigation, hunting, and schooling. The receptors of the electrical sense are modified hair cells of the lateral line system.

Fish and some aquatic amphibians detect hydrodynamic stimuli via a lateral line. This system consists of an array of sensors called neuromasts along the length of the fish's body. Neuromasts can be free-standing (superficial neuromasts) or within fluid-filled canals (canal neuromasts). The sensory cells within neuromasts are polarized hair cells contained within a gelatinous cupula. The cupula, and the stereocilia within, are moved by a certain amount depending on the movement of the surrounding water. Afferent nerve fibers are excited or inhibited depending on whether the hair cells they arise from are deflected in the preferred or opposite direction. Lateral line receptors form somatotopic maps within the brain informing the fish of amplitude and direction of flow at different points along the body. These maps are located in the medial octavolateral nucleus (MON) of the medulla and in higher areas such as the torus semicircularis.

Pressure Detection

Pressure detection uses the organ of Weber, a system consisting of three appendages of vertebrae transferring changes in shape of the gas bladder to the middle ear. It can be used to regulate the buoyancy of the fish. Fish like the weather fish and other loaches are also known to respond to low pressure areas but they lack a swim bladder.

Chemoreception

The shape of the hammerhead shark's head may enhance olfaction by spacing the nostrils further apart.

Sharks have keen olfactory senses, located in the short duct (which is not fused, unlike bony fish) between the anterior and posterior nasal openings, with some species able to detect as little as one part per million of blood in seawater.

Sharks have the ability to determine the direction of a given scent based on the timing of scent detection in each nostril. This is similar to the method mammals use to determine direction of sound.

They are more attracted to the chemicals found in the intestines of many species, and as a result often linger near or in sewage outfalls. Some species, such as nurse sharks, have external barbels that greatly increase their ability to sense prey.

The MHC genes are a group of genes present in many animals and important for the immune system; in general, offspring from parents with differing MHC genes have a stronger immune system. Fish are able to smell some aspect of the MHC genes of potential sex partners and prefer partners with MHC genes different from their own.

Salmon have a strong sense of smell. Speculation about whether odours provide homing cues, go back to the 19th century. In 1951, Hasler hypothesised that, once in vicinity of the estuary or entrance to its birth river, salmon may use chemical cues which they can smell, and which are unique to their natal stream, as a mechanism to home onto the entrance of the stream. In 1978, Hasler and his students convincingly showed that the way salmon locate their home rivers with such precision was indeed because they could recognise its characteristic smell. They further demonstrated that the smell of their river becomes imprinted in salmon when they transform into smolts, just before they migrate out to sea. Homecoming salmon can also recognise characteristic smells in tributary streams as they move up the main river. They may also be sensitive to characteristic pheromones given off by juvenile conspecifics. There is evidence that they can "discriminate between two populations of their own species".

Electroreception and Magnetoreception

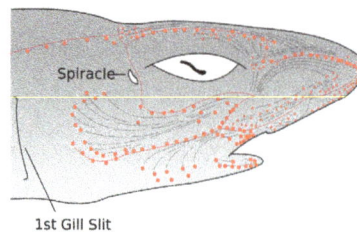

Electromagnetic field receptors (ampullae of Lorenzini) and motion detecting canals in the head of a shark

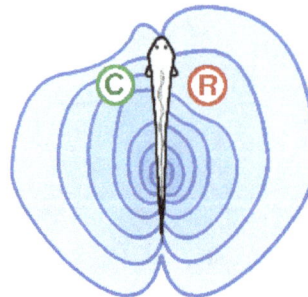

Active electrolocation. Conductive objects concentrate the field and resistive objects spread the field.

Electroreception, or electroception, is the ability to detect electric fields or currents. Some fish, such as catfish and sharks, have organs that detect weak electric potentials on the order of millivolts. Other fish, like the South American electric fishes Gymnotiformes, can produce weak electric currents, which they use in navigation and social communication. In sharks, the ampullae of Lorenzini are electroreceptor organs. They number in the hundreds to thousands. Sharks use the ampullae of Lorenzini to detect the electromagnetic fields that all living things produce. This helps sharks (particularly the hammerhead shark) find prey. The shark has the greatest electrical sensitivity of any animal. Sharks find prey hidden in sand by detecting the electric fields they produce. Ocean currents moving in the magnetic field of the Earth also generate electric fields that sharks can use for orientation and possibly navigation.

Electric field proximity sensing is used by the electric catfish to navigate through muddy waters. These fish make use of spectral changes and amplitude modulation to determine factors such shape, size, distance, velocity, and conductivity. The abilities of the electric fish to communicate and identify sex, age, and hierarchy within the species are also made possible through electric fields. EF gradients as low as 5nV/cm can be found in some saltwater weakly electric fish.

The paddlefish (*Polyodon spathula*) hunts plankton using thousands of tiny passive electroreceptors located on its extended snout, or rostrum. The paddlefish is able to detect electric fields that oscillate at 0.5–20 Hz, and large groups of plankton generate this type of signal.

Electric fishes use an active sensory system to probe the environment and create active electrodynamic imaging.

In 1973, it was shown that Atlantic salmon have conditioned cardiac responses to electric fields with strengths similar to those found in oceans. "This sensitivity might allow a migrating fish to align itself upstream or downstream in an ocean current in the absence of fixed references."

Magnetoception, or magnetoreception, is the ability to detect the direction one is facing based on the Earth's magnetic field. In 1988, researchers found iron, in the form of single domain magnetite, resides in the skulls of sockeye salmon. The quantities present are sufficient for magnetoception.

Fish Navigation

Salmon regularly migrate thousands of miles to and from their breeding grounds.

Salmon spend their early life in rivers, and then swim out to sea where they live their adult lives and gain most of their body mass. When they have matured, they return to the rivers to spawn. Usually they return with uncanny precision to the natal river where they were born, and even to the very spawning ground of their birth. It is thought that, when they are in the ocean, they use magnetoception to locate the general position of their natal river, and once close to the river, that they use their sense of smell to home in on the river entrance and even their natal spawning ground.

After several years wandering huge distances in the ocean, most surviving salmon return to the same natal rivers where they were spawned. Then most of them swim up the rivers until they reach the very spawning ground that was their original birthplace.

There are various theories about how this happens. One theory is that there are geomagnetic and chemical cues which the salmon use to guide them back to their birthplace. The fish may be sensitive to the Earth's magnetic field, which could allow the fish to orient itself in the ocean, so it can navigate back to the estuary of its natal stream.

Pain

Experiments done by William Tavolga provide evidence that fish have pain and fear responses. For instance, in Tavolga's experiments, toadfish grunted when electrically shocked and over time they came to grunt at the mere sight of an electrode.

Hooked sailfish

In 2003, Scottish scientists at the University of Edinburgh and the Roslin Institute concluded that rainbow trout exhibit behaviors often associated with pain in other animals. Bee venom and acetic acid injected into the lips resulted in fish rocking their bodies and rubbing their lips along the sides and floors of their tanks, which the researchers concluded were attempts to relieve pain, similar to what mammals would do. Neurons fired in a pattern resembling human neuronal patterns.

Professor James D. Rose of the University of Wyoming claimed the study was flawed since it did not provide proof that fish possess "conscious awareness, particularly a kind of awareness that is meaningfully like ours". Rose argues that since fish brains are so different from human brains, fish are probably not conscious in the manner humans are, so that reactions similar to human reactions to pain instead have other causes. Rose had published a study a year earlier arguing that fish cannot feel pain because their brains lack a neocortex. However, animal behaviorist Temple Grandin argues that fish could still have consciousness without a neocortex because "different species can use different brain structures and systems to handle the same functions."

Animal welfare advocates raise concerns about the possible suffering of fish caused by angling. Some countries, such as Germany have banned specific types of fishing, and the British RSPCA now prosecutes individuals who are cruel to fish.

Pain in Fish

Whether fish, such as this hooked salmon, can be said to feel pain is controversial

Pain in fish is a contentious issue. Pain is a complex mental state, with a distinct perceptual quality but also associated with suffering, which is an emotional state. Because of this complexity, the presence of pain in an animal, or another human for that matter, cannot be determined unambiguously using observational methods, but the conclusion that animals experience pain is often inferred on the basis of likely presence of phenomenal consciousness which is deduced from comparative brain physiology as well as physical and behavioural reactions.

Fish fulfill several criteria proposed as indicating that non-human animals may experience pain. These fulfilled criteria include a suitable nervous system and sensory receptors, opioid receptors and reduced responses to noxious stimuli when given analgesics and local anaesthetics, physiological changes to noxious stimuli, displaying protective motor reactions, exhibiting avoidance learning and making trade-offs between noxious stimulus avoidance and other motivational requirements.

If fish feel pain, there are ethical and animal welfare implications including the consequences of exposure to pollutants, and practices involving commercial and recreational fishing, aquaculture, in ornamental fish and genetically modified fish and for fish used in scientific research.

Background

The possibility that fish and other non-human animals may experience pain has a long history. Initially, this was based around theoretical and philosophical argument, but more recently has turned to scientific investigation.

Philosophy

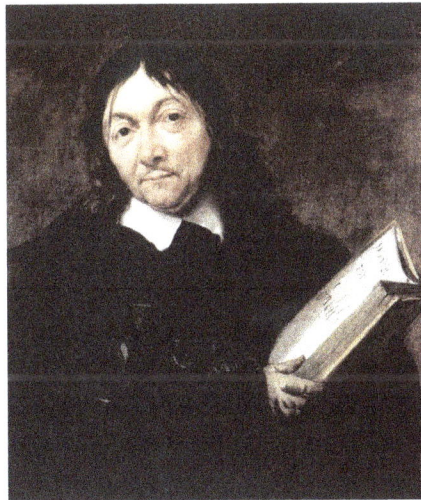

René Descartes

The idea that non-human animals might not feel pain goes back to the 17th-century French philosopher, René Descartes, who argued that animals do not experience pain and suffering because they lack consciousness. In 1789, the British philosopher and social reformist, Jeremy Bentham, addressed in his book *An Introduction to the Principles of Morals and Legislation* the issue of our treatment of animals with the following often quoted words: "The question is not, Can they reason? nor, can they talk? but, Can they suffer?"

Peter Singer, a bioethicist and author of *Animal Liberation* published in 1975, suggested that consciousness is not necessarily the key issue: just because animals have smaller brains, or are 'less conscious' than humans, does not mean that they are not capable of feeling pain. He goes on further to argue that we do not assume newborn infants, people suffering from neurodegenerative brain diseases or people with learning disabilities experience less pain than we would.

Bernard Rollin, the principal author of two U.S. federal laws regulating pain relief for animals, writes that researchers remained unsure into the 1980s as to whether animals experience pain, and veterinarians trained in the U.S. before 1989 were taught to simply ignore animal pain. In his interactions with scientists and other veterinarians, Rollin was regularly asked to "prove" that animals are conscious, and to provide "scientifically acceptable" grounds for claiming that they feel pain.

Continuing into the 1990s, discussions were further developed on the roles that philosophy and science had in understanding animal cognition and mentality. In subsequent years, it was argued there was strong support for the suggestion that some animals (most likely amniotes) have at least simple conscious thoughts and feelings and that the view animals feel pain differently to humans is now a minority view.

Scientific Investigation

Cambridge Declaration on Consciousness (2012)

The absence of a neocortex does not appear to preclude an organism from experiencing affective states. Convergent evidence indicates that non-human animals have the neuroanatomical, neurochemical, and neurophysiological substrates of conscious states along with the capacity to exhibit intentional behaviors. Consequently, the weight of evidence indicates that humans are not unique in possessing the neurological substrates that generate consciousness. Non-human animals, including all mammals and birds, and many other creatures, including octopuses, also possess these neurological substrates.

In the 20th and 21st centuries, there were many scientific investigations of pain in non-human animals.

Mammals

At the turn of the century, studies were published showing that arthritic rats self-select analgesic opiates. In 2014, the veterinary *Journal of Small Animal Practice* published an article on the recognition of pain which started – "The ability to experience pain is universally shared by all mammals..." and in 2015, it was reported in the science journal *Pain*, that several mammalian species (rat, mouse, rabbit, cat and horse) adopt a facial expression in response to a noxious stimulus that is consistent with the expression of humans in pain.

Birds

At the same time as the investigations using arthritic rats, studies were published showing that birds with gait abnormalities self-select for a diet that contains carprofen, a human analgesic. In 2005, it was written "Avian pain is likely analogous to pain experienced by most mammals" and in 2014, "...it is accepted that birds perceive and respond to noxious stimuli and that birds feel pain".

Reptiles and Amphibians

Veterinary articles have been published stating both reptiles and amphibians experience pain in a way analogous to humans, and that analgesics are effective in these two classes of vertebrates.

Argument by Analogy

In 2012 the American philosopher Gary Varner reviewed the research literature on pain in animals. His findings are summarised in the following table.

Argument by analogy					
Property	**Fish**	**Amphibians**	**Reptiles**	**Birds**	**Mammals**
Has nociceptors	✓	✓	✓	✓	✓
Has brain	✓	✓	✓	✓	✓
Nociceptors and brain linked	✓	✓	✓	✓	✓
Has endogenous opioids	✓	✓	✓	✓	✓
Analgesics affect responses	✓			✓	✓
Response to damaging stimuli similar to humans	✓	✓	✓	✓	✓

Arguing by analogy, Varner claims that any animal which exhibits the properties listed in the table could be said to experience pain. On that basis, he concludes that all vertebrates, including fish, probably experience pain, but invertebrates apart from cephalopods probably do not experience pain.

The Experience of Pain

Although there are numerous definitions of pain, almost all involve two key components.

First, nociception is required. This is the ability to detect noxious stimuli which evoke a reflex response that rapidly moves the entire animal, or the affected part of its body, away from the source of the stimulus. The concept of nociception does not imply any adverse, subjective "feeling" – it is a reflex action. An example in humans would be the rapid withdrawal of a finger that has touched something hot – the withdrawal occurs before any sensation of pain is actually experienced.

The second component is the experience of "pain" itself, or suffering – the internal, emotional interpretation of the nociceptive experience. Again in humans, this is when the withdrawn finger begins to hurt, moments after the withdrawal. Pain is therefore a private, emotional experience. Pain cannot be directly measured in other animals, including other humans; responses to putatively painful stimuli can be measured, but not the experience itself. To address this problem when assessing the capacity of other species to experience pain, argument-by-analogy is used. This is based on the principle that if an animal responds to a stimulus in a similar way to ourselves, it is likely to have had an analogous experience.

Nociception

Nociception usually involves the transmission of a signal along a chain of nerve fibers from the site of a noxious stimulus at the periphery to the spinal cord and brain. This process evokes a reflex arc response generated at the spinal cord and not involving the brain, such as flinching or withdrawal of a limb. Nociception is found, in one form or another, across all major animal taxa. Nociception can be observed using modern imaging techniques; and a physiological and behavioral response to nociception can often be detected. However, nociceptive responses can be so subtle in prey animals that trained (human) observers can not perceive them, whereas natural predators can and subsequently target injured individuals.

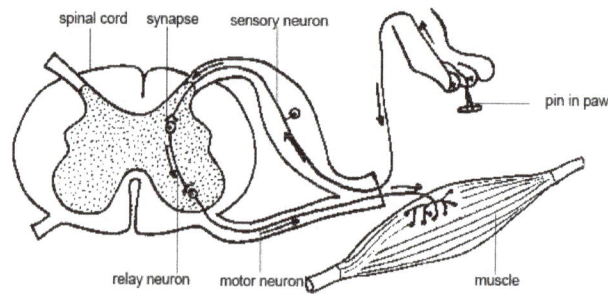

Nociception: The reflex arc of a dog with a pin in her paw. Note there is no communication to the brain, but the paw is withdrawn by nervous impulses generated by the spinal cord. There is no conscious interpretation of the stimulus by the dog.

Emotional Pain

Sometimes a distinction is made between "physical pain" and "emotional" or "psychological pain". Emotional pain is the pain experienced in the absence of physical trauma, e.g. the pain experienced by humans after the loss of a loved one, or the break-up of a relationship. It has been argued that only primates and humans can feel "emotional pain", because they are the only animals that have a neocortex – a part of the brain's cortex considered to be the "thinking area". However, research has provided evidence that monkeys, dogs, cats and birds can show signs of emotional pain and display behaviours associated with depression during painful experience, i.e. lack of motivation, lethargy, anorexia, unresponsiveness to other animals.

Physical Pain

The nerve impulses of the nociception response may be conducted to the brain thereby registering the location, intensity, quality and unpleasantness of the stimulus. This subjective component of pain involves conscious awareness of both the sensation and the unpleasantness (the aversive, negative affect). The brain processes underlying conscious awareness of the unpleasantness (suffering), are not well understood.

There have been several published lists of criteria for establishing whether non-human animals experience pain, e.g. Some criteria that may indicate the potential of another species, including fishes, to feel pain include:

1.   Has a suitable nervous system and sensory receptors

2.   Has opioid receptors and shows reduced responses to noxious stimuli when given analgesics and local anaesthetics

3. Physiological changes to noxious stimuli

4. Displays protective motor reactions that might include reduced use of an affected area such as limping, rubbing, holding or autotomy

5. Shows avoidance learning

6. Shows trade-offs between noxious stimulus avoidance and other motivational requirements

7. High cognitive ability and sentience

Adaptive Value

The adaptive value of nociception is obvious; an organism detecting a noxious stimulus immediately withdraws the limb, appendage or entire body from the noxious stimulus and thereby avoids further (potential) injury. However, a characteristic of pain (in mammals at least) is that pain can result in hyperalgesia (a heightened sensitivity to noxious stimuli) and allodynia (a heightened sensitivity to non-noxious stimuli). When this heightened sensitisation occurs, the adaptive value is less clear. First, the pain arising from the heightened sensitisation can be disproportionate to the actual tissue damage caused. Second, the heightened sensitisation may also become chronic, persisting well beyond the tissues healing. This can mean that rather than the actual tissue damage causing pain, it is the pain due to the heightened sensitisation that becomes the concern. This means the sensitisation process is sometimes termed maladaptive. It is often suggested hyperalgesia and allodynia assist organisms to protect themselves during healing, but experimental evidence to support this has been lacking.

In 2014, the adaptive value of sensitisation due to injury was tested using the predatory interactions between longfin inshore squid (*Doryteuthis pealeii*) and black sea bass (*Centropristis striata*) which are natural predators of this squid. If injured squid are targeted by a bass, they began their defensive behaviours sooner (indicated by greater alert distances and longer flight initiation distances) than uninjured squid. If anaesthetic (1% ethanol and $MgCl_2$) is administered prior to the injury, this prevents the sensitisation and blocks the behavioural effect. The authors claim this study is the first experimental evidence to support the argument that nociceptive sensitisation is actually an adaptive response to injuries.

The question has been asked, "If fish cannot feel pain, why do stingrays have purely defensive tail spines that deliver venom? Stingrays' ancestral predators are fish. And why do many fishes possess defensive fin spines, some also with venom that produces pain in humans?"

Peripheral Nervous System

Receptors

Primitive fish such as lampreys (*Petromyzon marinus*) have free nerve endings in the skin that respond to heat and mechanical pressure. However, behavioural reactions associated with nociception have not been recorded and it is also difficult to determine whether the mechanoreceptors in lamprey are truly nociceptive-specific or simply pressure-specific.

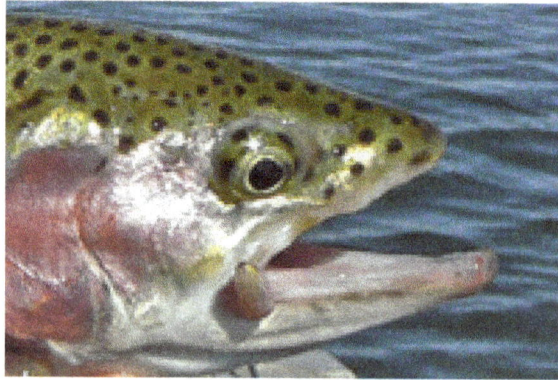
Rainbow trout have nociceptors on the face, snout and other areas of the body

Rainbow trout (*Oncorhynchus mykiss*) have polymodal nociceptors on the face and snout which respond to mechanical pressure, temperatures in the noxious range (> 40°C) and 1% acetic acid (a chemical irritant). Further studies found nociceptors to be more widely distributed over the bodies of rainbow trout, and also those of cod and carp. The most sensitive areas of the body are around the eyes, nostrils, fleshy parts of the tail and the pectoral and dorsal fins.

Rainbow trout also have corneal nociceptors. Out of 27 receptors investigated in one study, seven were polymodal nociceptors and six were mechanothermal nociceptors. Mechanical and thermal thresholds were lower than those of cutaneous receptors, indicating greater sensitivity in the cornea.

Bony fish possess nociceptors that are similar in function to those in mammals.

Nerve Fibres

There are two types of nerve fibre relevant to pain in fish. Group C nerve fibres are a type of sensory nerve fibre which lack a myelin sheath and have a small diameter, meaning they have a low nerve conduction velocity. The suffering that humans associate with burns, toothaches, or crushing injury are caused by C fibre activity. A typical human cutaneous nerve contains 83% Group C nerve fibres. A-delta fibres are another type of sensory nerve fibre, however, these are myelinated and therefore transmit impulses faster than non-myelinated C fibres. A-delta fibres carry cold, pressure and some pain signals, and are associated with acute pain that results in "pulling away" from noxious stimuli.

Bony fish possess both Group C and A-delta fibres representing 38.7% (combined) of the fibres in the tail nerves of common carp and 36% of the trigeminal nerve of rainbow trout. However, only 5% and 4% of these are C fibres in the carp and rainbow trout, respectively.

Some species of cartilagenous fish possess A-delta fibres, however, C fibres are either absent or found in very low numbers. The Agnatha (hagfishes and lamprey) primarily have Group C fibres.

Central Nervous System

The central nervous system (CNS) of fish contains a spinal cord, medulla oblongata, and the brain, divided into telencephalon, diencephalon, mesencephalon and cerebellum.

In fish, similar to other vertebrates, nociception travels from the peripheral nerves along the spinal nerves and is relayed through the spinal cord to the thalamus. The thalamus is connected to the telencephalon by multiple connections through the grey matter pallium, which has been demonstrated to receive nerve relays for noxious and mechanical stimuli.

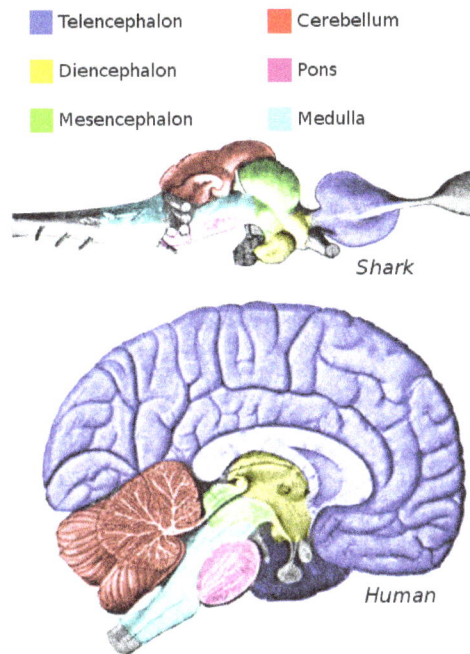

The brain regions of sharks and humans

The major tracts that convey pain information from the periphery to the brain are the spinothalamic tract (body) and the trigeminal tract (head). Both have been studied in agnathans, teleost, and elasmobranch fish (trigeminal in the common carp, spinothalamic tract in the sea robin, *Prionotus carolinus*).

Brain

If sensory responses in fish are limited to the spinal cord and hindbrain, they might be considered as simply reflexive. However, recordings from the spinal cord, cerebellum, tectum and telencephalon in both trout and goldfish (*Carassius auratus*) show these all respond to noxious stimuli. This indicates a nociceptive pathway from the periphery to the higher CNS of fish.

Microarray analysis of gene expression shows the brain is active at the molecular level in the forebrain, midbrain and hindbrain of common carp and rainbow trout. Several genes involved in mammalian nociception, such as brain-derived neurotrophic factor (BDNF) and the cannabinoid CB1 receptor are regulated in the fish brain after a nociceptive event.

Somatosensory evoked potentials (SEPs) are weak electric responses in the CNS following stimulation of peripheral sensory nerves. These further indicate there is a pathway from the peripheral nociceptors to higher brain regions. In goldfish, rainbow trout, Atlantic salmon (*Salmo salar*) and Atlantic cod (*Gadus morhua*), it has been demonstrated that putatively non-noxious and noxious stimulation elicit SEPs in different brain regions, including the telencephalon which may mediate the co-ordination of pain information. Moreover, multiple functional magnetic resonance imaging (fMRI) studies with several species of fishes have shown that when suffering from putative pain, there is profound activity in the forebrain which is highly reminiscent of that observed in humans and would be taken as evidence of the experience of pain in mammals.

Therefore, "higher" brain areas are activated at the molecular, physiological, and functional levels in fish experiencing a potentially painful event. Sneddon stated "This gives much weight to the proposal that fish experience some form of pain rather than a nociceptive event".

Opioid System and Effects of Analgesics

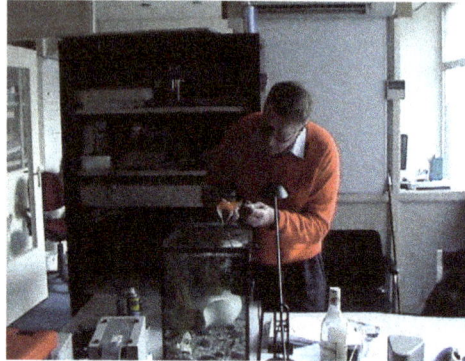

Analgesics and anaesthetics are commonly used for surgery on fish.

Teleost fish have a functional opioid system which includes the presence of opioid receptors similar to those of mammals. All four of the main opioid receptor types (delta, kappa, mu, and NOP) are conserved in vertebrates, even in primitive jawless fishes (agnathastoma).

The same analgesics and anaesthetics used in humans and other mammals, are often used for fish in veterinary medicine. These chemicals act on the nociceptive pathways, blocking signals to the brain where emotional responses to the signals are further processed by certain parts of the brain found in amniotes ("higher vertebrates").

Effects of Morphine

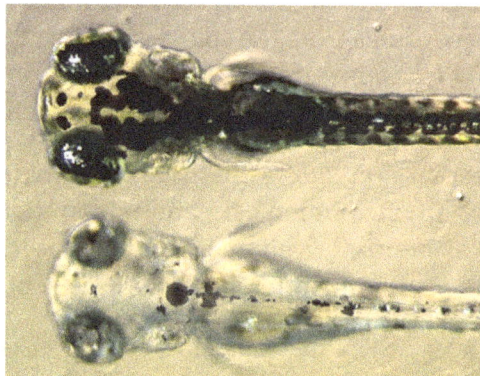

Five-day-old zebrafish larvae show behavioural responses indicative of pain in response to diluted acetic acid.

Pre-treatment with morphine (an analgesic in humans and other mammals) has a dose-dependent anti-nociceptive effect and mitigates the behavioural and ventilation rate responses of rainbow trout to noxious stimuli.

When acetic acid is injected into the lips of rainbow trout, they exhibit anomalous behaviours such as side-to-side rocking and rubbing their lips along the sides and floors of the tanks, and their ventilation rate increases. Injections of morphine reduce both the anomalous, noxious-stimulus related behaviours and the increase in ventilation rate. When the same noxious stimulus is applied to zebrafish (*Danio rerio*), they respond by decreasing their activity. As with the rainbow trout, morphine injected prior to the acid injection attenuates the decrease in activity in a dose-dependent manner.

Injection of acetic acid into the lips of rainbow trout causes a reduction in their natural neophobia (fear of novelty); this is reversed by the administration of morphine.

In goldfish injected with morphine or saline and then exposed to unpleasant temperatures, fish injected with saline acted with defensive behaviours indicating anxiety, wariness and fear, whereas those given morphine did not.

Effects of other Analgesics

The neurotransmitter, Substance P and the analgesic opioid enkephalins and β-endorphin, which act as endogenous analgesics in mammals, are present in fish.

Different analgesics have different effects on fish. In a study on the efficacy of three types of analgesic, buprenorphine (an opioid), carprofen (a non-steroidal anti-inflammatory drug) and lidocaine (a local anaesthetic), ventilation rate and time to resume feeding were used as pain indicators. Buprenorphine had limited impact on the fish's response, carprofen ameliorated the effects of noxious stimulation on time to resume feeding, however, lidocaine reduced all the behavioural indicators.

Tramadol also increases the nociceptive threshold in fish, providing further evidence of an anti-nociceptive opioid system in fish.

Effects of Naloxone

Naloxone is an μ-opioid receptor antagonist which, in mammals, negates the analgesic effects of opioids. Both adult and 5-day old zebrafish larvae show behavioural responses indicative of pain in response to injected or diluted acetic acid. The anti-nociceptive properties of morphine or buprenorphine are reversed if adults, or larvae, are co-treated with naloxone. Both naloxone and prolyl-leucyl-glycinamide (another opiate antagonist in mammals) reduced the analgesic effects of morphine to electric shocks received by goldfish, indicating they can act as an opiate antagonist in fish.

Physiological Changes

The physiological changes of fish in response to noxious stimuli include elevations of ventilation rate and cortisol levels.

Protective Responses

Studies show that fish exhibit protective behavioural responses to putatively painful stimuli.

Noxiously stimulated common carp show anomalous rocking behaviour and rub their lips against the tank walls.

Noxiously stimulated zebrafish reduce their frequency of swimming and increase their ventilation rate.

Noxiously stimulated Atlantic cod display increased hovering close to the bottom of the tank and reduced use of shelter.

When acetic acid or bee venom is injected into the lips of rainbow trout, they exhibit an anomalous side-to-side rocking behaviour on their pectoral fins, rub their lips along the sides and floors of the tanks and increase their ventilation rate. When acetic acid is injected into the lips of zebrafish, they respond by decreasing their activity. The magnitude of this behavioural response depends on the concentration of the acetic acid.

The behavioural responses to a noxious stimulus differ between species of fish. Noxiously stim- ulated common carp (*Cyprinus carpio*) show anomalous rocking behaviour and rub their lips against the tank walls, but do not change other behaviours or their ventilation rate. In contrast, zebrafish (*Danio rerio*) reduce their frequency of swimming and increase their ventilation rate but do not display anomalous behaviour. Rainbow trout, like the zebrafish, reduce their frequency of swimming and increase their ventilation rate. Nile tilapia (*Oreochromis niloticus*), in response to a tail fin clip, increase their swimming activity and spend more time in the light area of their tank.

Since this initial work, Sneddon and her co-workers have shown that rainbow trout, common carp and zebrafish experiencing a noxious stimulation exhibit rapid changes in physiology and behavior that persist for up to 6 hours and thus are not simple reflexes.

Five-day old zebrafish larvae show a concentration dependent increase in locomotor activity in response to different concentrations of diluted acetic acid. This increase in locomotor activity is accompanied by an increase in cox-2 mRNA, demonstrating that nociceptive pathways are also activated.

Fish show different responses to different noxious stimuli, even when these are apparently similar. This indicates the response is flexible and not simply a nociceptive reflex. Atlantic cod injected in the lip with acetic acid, capsaicin, or piercing the lip with a commercial fishing hook, showed different responses to these three types of noxious stimulation. Those cod treated with acetic acid and capsaicin displayed increased hovering close to the bottom of the tank and reduced use of shelter. However, hooked cod only showed brief episodes of head shaking.

Avoidance Learning

Early experiments provided evidence that fish learn to respond to putatively noxious stimuli. For instance, toadfish (*Batrachoididae*) grunt when they are electrically shocked, but after repeated shocks, they grunt simply at the sight of the electrode. More recent studies show that both goldfish and trout learn to avoid locations in which they receive electric shocks. Furthermore, this avoidance learning is flexible and is related to the intensity of the stimulus.

Trade-offs in Motivation

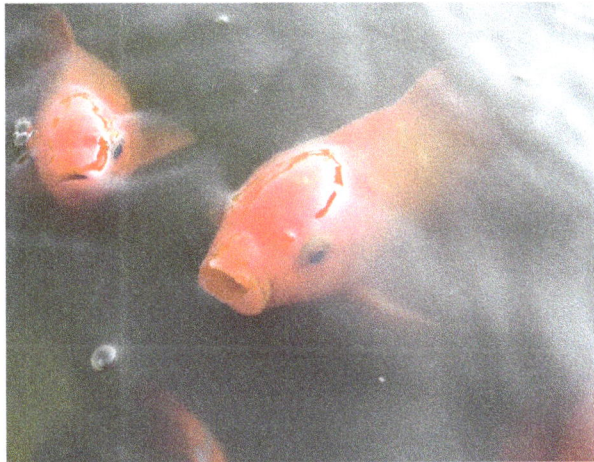
Goldfish make trade-offs between their motivation to feed or avoid an acute noxious stimulus.

A painful experience may change the motivation for normal behavioural responses.

In a 2007 study, goldfish were trained to feed at a location of the aquarium where subsequently they would receive an electric shock. The number of feeding attempts and time spent in the feeding/shock zone decreased with increased shock intensity and with increased food deprivation the number and the duration of feeding attempts increased as did escape responses as this zone was entered. The researchers suggested that goldfish make a trade-off in their motivation to feed with their motivation to avoid an acute noxious stimulus.

Rainbow trout naturally avoid novelty (i.e. they are neophobic). Braithwaite describes a study in which a brightly coloured Lego brick is placed in the tank of rainbow trout. Trout injected in the lip with a small amount of saline strongly avoided the Lego brick, however, trout injected with acetic acid spent considerably more time near the Lego block. When the study was repeated but with the fish also being given morphine, the avoidance response returned in those fish injected with acetic acid and could not be distinguished from the responses of saline injected fish.

To explore the possibility of a trade-off between responding to a noxious stimulus and predation, researchers presented rainbow trout with a competing stimulus, a predator cue. Noxiously stimulated fish cease showing anti-predator responses, indicating that pain becomes their primary motivation. The same study investigated the potential trade-off between responding to a noxious stimulus and social status. The responses of the noxiously treated trout varied depending on the familiarity of the fish they were placed with. The researchers suggested the findings of the motivational changes and trade-offs provide evidence for central processing of pain rather than merely showing a nociceptive reflex.

Paying a Cost for Analgesia

Zebrafish given access to a barren, brightly lit chamber or an enriched chamber prefer the enriched area. When these fish are injected with acetic acid or saline as a control they still choose the same enriched chamber. However, if an analgesic is dissolved in the barren, less-preferred chamber, zebrafish injected with noxious acid lose their preference and spend over half their time in the previously less-favourable, barren chamber. This suggests a trade-off in motivation and furthermore, they are willing to pay a cost to enter a less preferred environment to access pain relief.

Cognitive Ability and Sentience

The learning abilities of fish demonstrated in a range of studies indicate sophisticated cognitive processes that are more complex than simple associative learning. Examples include the ability to recognise social companions, avoidance (for some months or years) of places where they encountered a predator or were caught on a hook and forming mental maps.

It has been argued that although a high cognitive capacity may indicate a greater likelihood of experiencing pain, it also gives these animals a greater ability to deal with this, leaving animals with a lower cognitive ability a greater problem in coping with pain.

Societal Implications

There are concerns that angling causes pain in fish.

Given the broad scientific consensus that fish can probably feel pain, it has been suggested that precautionary principles should be applied to commercial fishing, which would likely have multiple consequences.

Both scientists and animal protection advocates have raised concerns about the possible suffering (pain and fear) of fish caused by angling.

Other societal implications of fish experiencing pain include acute and chronic exposure to pollutants, commercial and sporting fisheries (e.g. injury during trawling, tagging/fin clipping during stock assessment, tissue damage, physical exhaustion and severe oxygen deficit during capture, pain and stress during slaughter, use of live bait), aquaculture (e.g. tagging/fin clipping, high stocking densities resulting in increased aggression, food deprivation for disease treatment or before harvest, removal from water for routine husbandry, pain during slaughter), ornamental fish (e.g. capture by sub-lethal poisoning, permanent adverse physical states due to selective breeding), scientific research (e.g. genetic-modification) may have detrimental effects on welfare, deliberately-imposed adverse physical, physiological and behavioural states, electrofishing, tagging, fin clipping or otherwise marking fish, handling procedures which may cause injury.

Legislation

In the UK, the legislation protecting animals during scientific research, the "Animals (Scientific Procedures) Act 1986", protects fish from the moment they become capable of independent feeding. The legislation protecting animals in most other circumstances in the UK is "The Animal Welfare Act, 2006" which states that in the Act, " "animal" means a vertebrate other than man", clearly including fish.

In the US, the legislation protecting animals during scientific research is "The Animal Welfare Act". This excludes protection of "cold-blooded" animals, including fish.

The 1974 Norwegian Animal Rights Law states it relates to mammals, birds, frogs, salamander, reptiles, fish, and crustaceans.

Controversy

Receptors and Nerve Fibres

It has been argued that fish can not feel pain because they do not have a sufficient density of appropriate nerve fibres. A typical human cutaneous nerve contains 83% Group C nerve fibres, however, the same nerves in humans with congenital insensitivity to pain have only 24-28% C type fibres. Based on this, James Rose, from the University of Wyoming, has argued that the absence of C type fibres in cartilagenous sharks and rays indicates that signalling leading to pain perception is likely to be impossible, and the low numbers for bony fish (e.g. 5% for carp and trout) indicate this is also highly unlikely for these fish. Rose concludes there is little evidence that sharks and rays possess the nociceptors required to initiate pain detection in the brain, and that, while bony fish are able to unconsciously learn to avoid injurious stimuli, they are little more likely to experience conscious pain than sharks.

A-delta type fibres, believed to trigger avoidance reactions, are common in bony fish, although they have not been found in sharks or rays.

Rose concludes that fishes have survived well in an evolutionary sense without the full range of nociception typical of humans or other mammals.

Brain

In 2002, Rose published reviews arguing that fish cannot feel pain because they lack a neocortex in the brain. (This would also rule out pain perception in most mammals, all birds and reptiles.) However, in 2003, a research team led by Lynne Sneddon concluded that the brains of rainbow trout fire neurons in the same way human brains do when experiencing pain. Rose criticized the study, claiming it was flawed, mainly because it did not provide proof that fish possess "conscious awareness, particularly a kind of awareness that is meaningfully like ours".

Rose, and more recently Brian Key from The University of Queensland, argue that because the fish brain is very different to humans, fish are probably not conscious in the manner humans are, and while fish may react in a way similar to the way humans react to pain, the reactions in the case of fish have other causes. Studies indicating that fish can feel pain were confusing nociception with feeling pain, says Rose. "Pain is predicated on awareness. The key issue is the distinction between nociception and pain. A person who is anaesthetised in an operating theatre will still respond physically to an external stimulus, but he or she will not feel pain." According to Rose and Key, the literature relating to the question of consciousness in fish is prone to anthropomorphisms and care is needed to avoid erroneously attributing human-like capabilities to fish. Sneddon suggests it is entirely possible that a species with a different evolutionary path could evolve different neural systems to perform the same functions (i.e. convergent evolution), as studies on the brains of birds have shown. Key agrees that phenomenal consciousness is likely to occur in mammals and birds, but not in fish. Animal behaviouralist Temple Grandin argues that fish could still have consciousness without a neocortex because "different species can use different brain structures and systems to handle the same functions." Sneddon proposes that to suggest a function suddenly arises without a primitive form defies the laws of evolution.

Other researchers also believe that animal consciousness does not require a neocortex, but can arise from homologous subcortical brain networks. It has been suggested that brainstem circuits can generate pain. This includes research with anencephalic children who, despite missing large portions of their cortex, express emotions. There is also evidence from activation studies showing brainstem mediated feelings in normal humans and foetal withdrawal responses to noxious stimulation but prior to development of the cortex.

Protective Responses

Initial work by Sneddon and her co-workers characterised behavioural responses in rainbow trout, common carp and zebrafish. However, when these experiments were repeated by Newby and Stevens without anaesthetic, rocking and rubbing behaviour was not observed, suggesting that some of the alleged pain responses observed by Sneddon and co-workers were likely to be due to recovery of the fish from anaesthesia.

Several researchers argue about the definition of pain used in behavioural studies, as the observations recorded were contradictory, non-validated and non-repeatable by other researchers. In 2012, Rose argued that fishes resume "normal feeding and activity immediately or soon after surgery".

Nordgreen said that the behavioural differences they found in response to uncomfortable temperatures showed that fish feel both reflexive and cognitive pain. "The experiment shows that fish do not only respond to painful stimuli with reflexes, but change their behavior also after the event," Nordgreen said. "Together with what we know from experiments carried out by other groups, this indicates that the fish consciously perceive the test situation as painful and switch to behaviors indicative of having been through an aversive experience." In 2012, Rose and others reviewed this and further studies which concluded that pain had been found in fish. They concluded that the results from such research are due to poor design and misinterpretation, and that the researchers were unable to distinguish unconscious detection of injurious stimuli (nociception) from conscious pain.

Electroreception

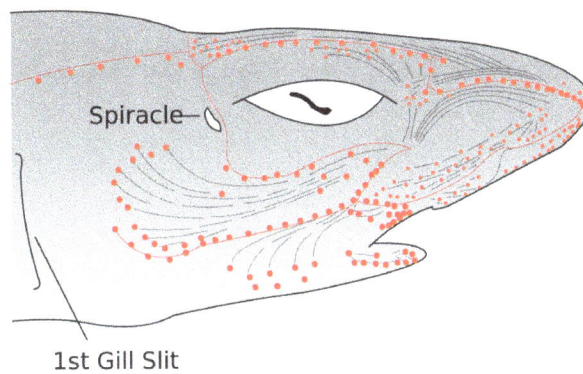

Electroreceptors (Ampullae of Lorenzini) and lateral line canals in the head of a shark.

Electroreception is the biological ability to perceive natural electrical stimuli. It has been observed almost exclusively in aquatic or amphibious animals, since salt-water is a much better conductor than air, the currently known exceptions being the platypus, echidnas, cockroaches and bees. Electroreception is used in electrolocation (detecting objects) and for electrocommunication.

Overview

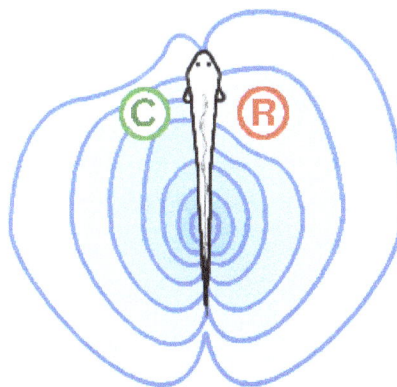

Active electrolocation. Conductive objects concentrate the field and resistive objects spread the field.

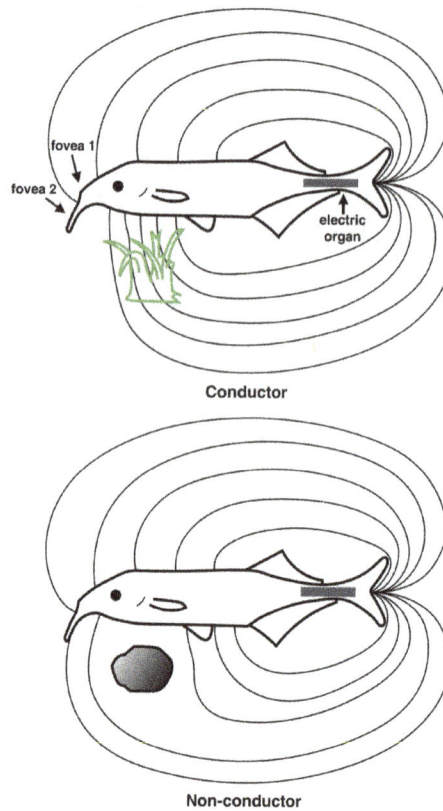

For the elephantfish (here *Gnathonemus*) the electric field emanates from an electric organ
in the tail region (gray rectangle). It is sensed by the electroreceptive skin areas, using
two electric pits (foveas) to actively search and inspect objects. Shown are
the field distortions created by two different types of objects: a plant
that conducts better than water, above (green) and
a non-conducting stone, below (gray).

Until recently, electroreception was known only in vertebrates. Recent research has shown that bees can detect the presence and pattern of a static charge on flowers. Electroreception is found in lampreys, cartilaginous fishes (sharks, rays, chimaeras), lungfishes, bichirs, coelacanths, sturgeons, paddlefishes, catfishes, gymnotiformes, elephantfishes, monotremes, and at least one species of cetacean. The electroreceptor organs in all these groups are derived embryologically from a mechanoreceptor system. In fishes they are developed from the lateral lines. In most groups electroreception is *passive*, where it is used predominantly in predation. Two groups of teleost fishes are weakly electric and engage in *active* electroreception; the Neotropical knifefishes (Gymnotiformes) and the African elephantfishes (Notopteroidei). A rare terrestrial exception is the Western long-beaked echidna which has about 2,000 electroreceptors on its bill, compared to 40,000 for its semi-aquatic monotreme relative, the duck-billed platypus.

Electrolocation

Electroreceptive animals use this sense to locate objects around them. This is important in ecological niches where the animal cannot depend on vision: for example in caves, in murky water and at night. Many fish use electric fields to detect buried prey. Some shark embryos and pups "freeze" when they detect the characteristic electric signal of their predators. It has been proposed that sharks can use their acute electric sense to detect the earth's magnetic field by detecting the weak electric currents induced by their swimming or by the flow of ocean currents. The walking behaviour of cockroaches can be affected by the presence of a static electric field: they like to avoid the electric field. Cabbage loopers are also known to avoid electric fields.

Active Electrolocation

In active electrolocation, the animal senses its surrounding environment by generating electric fields and detecting distortions in these fields using electroreceptor organs. This electric field is generated by means of a specialised electric organ consisting of modified muscle or nerves. This field may be modulated so that its frequency and wave form are unique to the species and sometimes, the individual (see Jamming avoidance response). Animals that use active electro-reception include the weakly electric fish, which either generate small electrical pulses (termed "pulse-type") or produce a quasi-sinusoidal discharge from the electric organ (termed "wave-type"). These fish create a potential which is usually smaller than one volt. Weakly electric fish can discriminate between objects with different resistance and capacitance values, which may help in identifying the object. Active electroreception typically has a range of about one body length, though objects with an electrical impedance similar to that of the surrounding water are nearly undetectable.

Passive Electrolocation

In passive electrolocation, the animal senses the weak bioelectric fields generated by other animals and uses it to locate them. These electric fields are generated by all animals due to the activity of their nerves and muscles. A second source of electric fields in fish is the ion pumps associated with osmoregulation at the gill membrane. This field is modulated by the opening and closing of the mouth and gill slits. Many fish that prey on electrogenic fish use the discharges of their prey to detect them. This has driven the prey to evolve more complex or higher frequency signals that are harder to detect.

Passive electroreception is carried out solely by ampullary electroreceptors in fish. It is tuned to low frequency signals (less than 1 Hz to tens of Hz).

Fish use passive electroreception to supplement or replace their other senses when detecting prey and predators. In sharks, sensing an electric dipole alone is sufficient to cause them to try to eat it.

Electrocommunication

Weakly electric fish can also communicate by modulating the electrical waveform they generate, an ability known as electrocommunication. They may use this for mate attraction and territorial displays. Some species of catfish use their electric discharges only in agonistic displays.

In one species of *Brachyhypopomus* (a genus of South American river fish belonging to the family Hypopomidae, commonly known as *bluntnose knifefishes*), the electric discharge pattern is similar to the low voltage electrolocative discharge of the electric eel. This is hypothesised to be a form of Batesian mimicry of the dangerous eel.

Sensory Mechanism

Active electroreception relies upon tuberous electroreceptors which are sensitive to high frequency (20-20,000 Hz) stimuli. These receptors have a loose plug of epithelial cells which capacitively couples the sensory receptor cells to the external environment. Passive electroreception however, relies upon ampullary receptors which are sensitive to low frequency stimuli (below 50 Hz). These receptors have a jelly-filled canal leading from the sensory receptors to the skin surface. Mormyrid electric fish from Africa use tuberous receptors known as Knollenorgans to sense electric communication signals.

Examples

Sharks and Rays

Sharks and rays (members of the subclass *Elasmobranchii*), such as the lemon shark, rely heavily on electrolocation in the final stages of their attacks, as can be demonstrated by the robust feeding response elicited by electric fields similar to those of their prey. Sharks are the most electrically sensitive animals known, responding to DC fields as low as 5 nV/cm.

The electric field sensors of sharks are called the ampullae of Lorenzini. They consist of electroreceptor cells connected to the seawater by pores on their snouts and other zones of the head. A problem with the early submarine telegraph cables was the damage caused by sharks who sensed the electric fields produced by these cables. It is possible that sharks may use Earth's magnetic field to navigate the oceans using this sense.

Bony Fish

The electric eel, besides its ability to generate high voltage electric shocks, uses lower voltage pulses for navigation and prey detection in its turbid habitat. This ability is shared with other gymnotiformes.

Monotremes

The monotremes are the only group of land mammals known to have evolved electroreception. While the electroreceptors in fish and amphibians evolved from mechanosensory lateral line organs, those of monotremes are based on cutaneous glands innervated by trigeminal nerves. The electroreceptors of monotremes consist of free nerve endings located in the mucous glands of the snout. Among the monotremes, the platypus (*Ornithorhynchus anatinus*) has the most acute electric sense. The platypus has almost 40,000 electroreceptors arranged in a series of stripes along the bill, which probably aids the localisation of prey. The platypus electroreceptive system is highly directional, with the axis of greatest sensitivity pointing outwards and downwards. By making short-latency head movements called "saccades" when swimming, platypuses constantly expose the most sensitive part of their bill to the stimulus to localise prey as accurately as possible. The platypus appears to use electroreception along with pressure sensors to determine the distance to prey from the delay between the arrival of electrical signals and pressure changes in the water.

The platypus also uses electroreception.

The electroreceptive capabilities of the two species of echidna (which are terrestrial) are much more simple. Long-beaked echidnas (genus *Zaglossus*) possess only 2,000 receptors and short-beaked echidnas (*Tachyglossus aculeatus*) merely 400 that are concentrated in the tip of the snout. This difference can be attributed to their habitat and feeding methods. Western long-beaked echidnas live in wet tropical forests where they feed on earthworms in damp leaf litter, so their habitat is probably favourable to the reception of electrical signals. Contrary to this is the varied but generally more arid habitat of their short-beaked relative which feeds primarily on termites and ants in nests; the humidity in these nests presumably allows electroreception to be used in hunting for buried prey, particularly after rains. Experiments have shown that echidnas can be trained to respond to weak electric fields in water and moist soil. The electric sense of the echidna is hypothesised to be an evolutionary remnant from a platypus-like ancestor.

Dolphins

Dolphins have evolved electroreception in structures different from those of fish, amphibians and monotremes. The hairless vibrissal crypts on the rostrum of the Guiana dolphin (*Sotalia guianensis*), originally associated with mammalian whiskers, are capable of electroreception as low as 4.8 μV/cm, sufficient to detect small fish. This is comparable to the sensitivity of electroreceptors in the platypus. To date (June 2013), these cells have been described from only a single dolphin specimen.

Bees

Bees collect a positive static charge while flying through the air (see Atmospheric electricity). When a bee visits a flower, the charge deposited on the flower takes a while to leak away into the ground. Bees can detect both the presence and the pattern of electric fields on flowers, and use this information to know if a flower has been recently visited by another bee and is therefore likely to have a reduced concentration of nectar. The mechanism of electric field reception in animals living in the air like bees is based on mechano- reception, not electroreception. Bees receive the electric field changes via the Johnston's organs in their antennae and possibly other mechano-receptors. They distinguish different temporal patterns and learn them. During the waggle dance, honeybees appear to use the electric field emanating from the dancing bee for distance communication.

Effects on Wildlife

It has been claimed that the electromagnetic fields generated by pylons and masts have adverse effects on wildlife; a list of 153 references to this has been published.

Fish Intelligence

The elephantnose fish has the highest brain-to-body oxygen consumption ratio of all known vertebrates

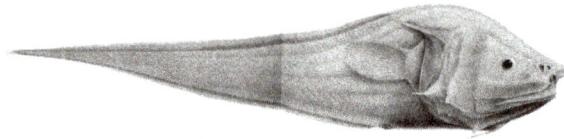

The bony-eared assfish has the smallest brain-to-body weight ratio of all known vertebrates

Fish intelligence is "...the resultant of the process of acquiring, storing in memory, retrieving, combining, comparing, and using in new contexts information and conceptual skills" as it applies to fish.

According to Culum Brown from Macquarie University, "Fish are more intelligent than they appear. In many areas, such as memory, their cognitive powers match or exceed those of 'higher' vertebrates including non-human primates."

Fish hold records for the relative brain weights of vertebrates. Most vertebrate species have similar brain-to-body mass ratios. The deep sea bathypelagic bony-eared assfish, has the smallest ratio of all known vertebrates. At the other extreme, the electrogenic elephantnose fish, an African freshwater fish, has one of the largest brain-to-body weight ratios of all known vertebrates (slightly higher than humans) and the highest brain-to-body oxygen consumption ratio of all known vertebrates (three times that for humans).

Brain

Fish typically have quite small brains relative to body size compared with other vertebrates, typically one-fifteenth the brain mass of a similarly sized bird or mammal. However, some fish have relatively large brains, most notably mormyrids and sharks, which have brains about as massive relative to body weight as birds and marsupials.

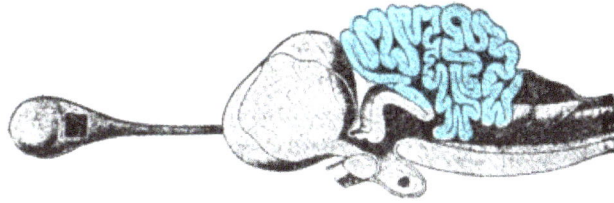

Cross-section of the brain of a porbeagle shark, with the cerebellum highlighted

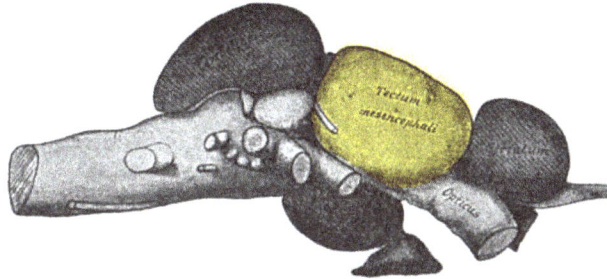

The brain of a cod

The cerebellum of cartilaginous and bony fishes is large and complex. In at least one important respect, it differs in internal structure from the mammalian cerebellum: The fish cerebellum does not contain discrete deep cerebellar nuclei. Instead, the primary targets of Purkinje cells are a distinct type of cell distributed across the cerebellar cortex, a type not seen in mammals. The circuits in the cerebellum are similar across all classes of vertebrates, including fish, reptiles, birds, and mammals. There is also an analogous brain structure in cephalopods with well-developed brains, such as octopuses. This has been taken as evidence that the cerebellum performs functions important to all animal species with a brain.

In mormyrid fish (a family of weakly electrosensitive freshwater fish), the cerebellum is considerably larger than the rest of the brain put together. The largest part of it is a special structure called the *valvula*, which has an unusually regular architecture and receives much of its input from the electrosensory system.

Memory

Red Sea clownfish can recognize their mate after 30 days separation.

Individual carp captured by anglers have been shown to become less catchable thereafter. This suggests that fish use their memory of negative experiences to associate capture with stress and therefore become less easy to catch. This type of associative learning has also been shown in paradise fish (*Macropodus opercularis*) which avoid places where they have experienced a single attack by a predator and continue to avoid for many months.

Some fish species can exhibit long-term memory. Anecdotally, channel catfish (*Ictalurus punctatus*) can remember the human voice call announcing food five years after last hearing that call. Goldfish remember the colour of a tube dispensing food one year after the last tube presentation. Sockeye salmon still react to a light signal that precedes food arrival up to eight months since the last reinforcement. Some common rudd and European chub could remember the person who trained them to feed from the hand, even after a 6-month break. Crimson-spotted rainbowfish can learn how to escape from a trawl by swimming through a small hole in the center and they remember this technique 11 months later. Rainbow trout can be trained to press a bar to get food, and they remember this three months after last seeing the bar. Red Sea clownfish can recognize their mate 30 days after it was experimentally removed from the home anemone.

Several fish species are capable of learning complex spatial relationships and forming cognitive maps. They can orient themselves using multiple landmarks or symbols and they are able to integrate experiences which enable them to generate appropriate avoidance responses.

Tool Use

Tool use is sometimes considered as an indication of intelligence in animals. There are few examples of tool use in fishes, perhaps because they have only their mouth in which to hold objects.

Several species of wrasse hold bivalves (scallops, clams and urchins) in their mouth and smash them against the surface of a rock (an "anvil") to break them up. This behaviour in an orange-dotted tuskfish (*Choerodon anchorago*) has been filmed; the fish fans sand to unearth the bivalve, takes it into its mouth, swims several metres to a rock which it uses as an anvil by smashing the mollusc apart with sideward thrashes of the head.

Archerfish (family Toxotidae) squirt jets of water at insects on plants above the surface to knock them into the water; they can adjust the size of the squirts to the size of the insect prey and learn to shoot at moving targets.

Whitetail damselfish clean the rock face where they intend to lay eggs by sucking up and blowing sand grains onto the surface. Triggerfish blow water at sea urchins to turn them over, thereby exposing their more vulnerable underside. Banded acaras (*Bujurquina vittata*) lay their eggs on a loose leaf and carry the leaf away when a predator approaches.

In one laboratory study, Atlantic cod (*Gadus morhua*) given access to an operant feeding machine learned to pull a string to get food. However, the researchers had also tagged the fish by threading a bead in front of their dorsal fin. Some fish snagged the string with their bead, resulting in food delivery. These fish eventually learned to swim in a particular way to repeatedly make the bead snag the string and get food. Because the fish used an object external to their body in a goal-oriented way, this satisfies some definitions of tool use.

Construction

As for tool use, construction behaviour may be mostly innate. Yet it can be sophisticated, and the fact that fish can make judicious repairs to their creation suggests intelligence. Construction methods in fishes can be divided into three categories: excavations, pile-ups, and gluing.

Excavations may be simple depressions dug up in the substrate, such as the nests of bowfin, smallmouth bass, and Pacific salmon, but it can also consist of fairly large burrows used for shelter and for nesting. Burrowing species include the mudskippers, the red band-fish *Cepola rubescens* (burrows up to 1 m deep, often with a side branch), the yellowhead jawfish *Opistognathus aurifrons* (chambers up to 22 cm deep, lined with coral fragments to solidify it), the convict blenny *Pholidichthys leucotaenia* whose burrow is a maze of tunnels and chambers thought to be as much as 6 m long, and the Nicaragua cichlid, *Hypsophrys nicaraguensis*, who drills a tunnel by spinning inside of it. In the case of the mudskippers, the burrows are shaped like a J and can be as much as 2 m deep. Two species, the giant mudskipper *Periophthalmodon schlosseri* and the walking goby *Scartelaos histophorus*, build a special chamber at the bottom of their burrows into which they carry mouthfuls of air. Once released the air accumulates at the top of the chamber and forms a reserve from which the fish can breathe – like all amphibious fishes, mudskippers are good air breathers. If researchers experimentally extract air from the special chambers, the fish diligently replenish it. The significance of this behaviour stems from the facts that at high tide, when water covers the mudflats, the fish stay in their burrow to avoid predators, and water inside the confined burrow is often poorly oxygenated. At such times these air-breathing fishes can tap into the air reserve of their special chambers .

Mounds are easy to build, but can be quite extensive. In North American streams, the male cutlip minnow *Exoglossum maxillingua*, 90–115 mm long (3.5-4.5 in), assembles mounds that are 75–150 mm high (3–6 in), 30–45 cm in diameter (12–18 in), made up of more than 300 pebbles 13–19 mm in diameter (a quarter to half an inch). The fish carry these pebbles one by one in their mouths, sometimes stealing some from the mounds of other males. The females deposit their eggs on the upstream slope of the mounds, and the males cover these eggs with more pebbles. Males of the hornyhead chub *Nocomis biguttatus*, 90 mm long (3.5 in), and of the river chub *Nocomis micropogon*, 100 mm long (4 in), also build mounds during the reproductive season. They start by clearing a slight depression in the substrate, which they overfill with up to 10,000 pebbles until the mounds are 60–90 cm (2–3 ft) long (in the direction of the water current), 30–90 cm wide (1–3 ft), and 5–15 cm high (2–6 in). Females lay their eggs among those pebbles. The stone accumulation is free of sand and it exposes the eggs to a good water current that supplies oxygen. Males of many mouthbrooding cichld species in Lake Malawi and Lake Tanganyika build sand cones that are flattened or crater-shaped at the top. Some of these mounds can be 3 m in diameter and 40 cm high. The mounds serve to impress females or to allow species recognition during courtship.

Male pufferfish, *Torquigener* sp., also build sand mounds to attract females. The mounds, up to 2 m in diameter, are intricate with radiating ridges and valleys.

Several species build up mounds of coral pieces either to protect the entrance to their burrows, as in tilefishes and gobies of the genus Valenciennea, or to protect the patch of sand in which they will bury themselves for the night, as in the Jordan's tuskfish *Choerodon jordani* and the rockmover wrasse *Novaculichthys taeniourus*.

Male sticklebacks are well known for their habit of building an enclosed nest made of pieces of vegetation glued together with secretions from their kidneys. Foam nests, made up of air bubbles glued together with mucus from the mouth, are also well known in gouramis and armoured catfish.

Social Intelligence

Fish can remember the attributes of other individuals, such as their competitive ability or past behavior, and modify their own behavior accordingly. For example, they can remember the identity of individuals to whom they have lost in a fight, and avoid these individuals in the future; or they can recognize territorial neighbors and show less aggression towards them as compared to strangers. They can recognize individuals in whose company they obtained less food in the past and preferentially associate with new partners in the future.

Fish can seem mindful of which individuals have watched them in the past. In an experiment with Siamese fighting fish, two males were made to fight each other while being watched by a female, whom the males could also see. The winner and the loser of the fight were then, separately, given a choice between spending time next to the watching female or to a new female. The winner courted both females equally, but the loser spent more time next to the new female, avoiding the watcher female. In this species, females prefer males they have seen win a fight over males they have seen losing, and it therefore makes sense for a male to prefer a female that has never seen him as opposed to a female that has seen him lose.

Knowing that if A>B and B>C, then A>C, is another type of evidence for intelligence, and it can be applied in the context of dominance hierarchies. In a study with the cichlid *Astatotilapia burtoni*, eight observer fish could watch individual A beat individual B, then B over C, C over D, and D over E. The observer fish were then given a choice of associating with either B or D (both of which they had seen win once and lose once). All eight observer fish spent more time next to D. Fish in this species prefer to associate with more subordinate individuals, so the preference for D showed that the observers had worked out that B was superior to C, and C to D, and therefore D was subordinate to B.

Deception

There are several examples of fish being deceptive, indicating they are capable of a theory of mind.

Distraction Display

In the threespine stickleback (*Gasterosteus aculeatus*), males sometimes see their nest full of eggs fall prey to groups of marauding females; some males, when they see a group of females approaching, swim away from their nest and start poking their snout into the substrate, as would a female raiding a nest. This distraction display commonly fools the females into behaving as if a nest has been discovered there and they rush to that site, leaving the male's real nest alone. Bowfin (*Amia calva*) males caring for their free-swimming fry exhibit a related distraction display when a potential fry-predator approaches; they move away and thrash about as if injured, drawing the predator's attention toward himself.

Adult male bowfins distract potential predators away from their fry by thrashing as if injured.

False Courtship Behaviour

In the Malili Lakes of Sulawesi, Indonesia, one species of sailfin silverside (*Telmatherina sarasinorum*) is an egg predator. They often follow courting pairs of the closely related species *T. antoniae*. When those pairs lay eggs, *T. sarasinorum* darts in and eats the eggs. On four different occasions in the field (out of 136 observation bouts in total), a male *T. sarasinorum* who was following a pair of courting *T. antoniae* eventually chased off the male *T. antoniae* and took his place, courting the heterospecific female. That female released eggs, at which point the male darted to the eggs and ate them.

Death Feigning

Death feigning as a way to attract prey is another form of deception. In Lake Malawi, the predatory cichlid *Nimbochromis livingstonii* have been seen first remaining stationary with their abdomen on or near sand and that then dropping onto their sides. In a variant behaviour, some *N. livingstonii* fell through the water column and landed onto their side. The fish then remained immobile for several minutes. Their colour pattern was blotchy and suggested a rotting carcass. Small inquisitive cichlids of other species often came near and they were suddenly attacked by the predator. About a third of the death-feigning performances led to an attack, and about one-sixth of the attacks were successful. Another African cichlid, *Lamprologus lemairii*, from Lake Tanganyika, has been reported to do the same thing. A South American cichlid, the yellowjacket cichlid *Parachromis friedrichsthalii*, also uses death feigning. They turn over onto their sides at the bottom of the sinkholes they inhabit and remain immobile for as long as 15 minutes, during which they attack the small mollies that come too close to them. The comb grouper *Mycteroperca acutirostris* may also be an actor, though in this case the behaviour should be called dying or illness feigning, rather than death feigning, because while lying on its side the fish occasionally undulates its body. In 1999, off the coast of southeastern Brazil, one juvenile comb grouper was observed using this tactic to catch five small prey in 15 minutes.

Cooperation

Cooperative foraging reflects some mental flexibility and planning, and could therefore be interpreted as intelligence. There are a few examples in fishes.

Yellowtail amberjack can form packs of 7-15 individuals that maneuver in U-shaped formations to cut away the tail end of prey shoals (jack mackerels or Cortez grunts) and herd the downsized shoal next to seawalls where they proceed to capture the prey.

In the coral reefs of the Red Sea, roving coralgrouper that have spotted a small prey fish hiding in a crevice sometimes visit the sleeping hole of a giant moray and shake their head at the moray, and this seems to be an invitation to group hunting as the moray often swims away with the grouper, is led to the crevice where the prey hides, and proceeds to probe that crevice (which is too small to let the grouper in), either catching the prey by itself or flushing it into the open where the grouper grabs it. The closely related coral trout also enrolls the help of moray eels in this way, and they only do so when the prey they seek is hidden in crevices, where only the eel can flush them. They also quickly learn to invite preferentially those individual eels that collaborate most often.

Similarly, zebra lionfish that detect the presence of small prey fishes flare up their fins as an invitation to other zebra lionfish, or even to another species of lionfish (*Pterois antennata*), to join them in better cornering the prey and taking turns at striking the prey so that every individual hunter ends up with similar capture rates.

Numeracy

Mosquitofish (*Gambusia holbrooki*) can distinguish between doors marked with either two or three geometric symbols, only one of which allows the fish to rejoin its shoalmates. This can be achieved when the two symbols have the same total surface area, density and brightness as the three symbols. Further studies show this discrimination extends to 4 vs 8, 15 vs 30, 100 vs 200, 7 vs 14, and 8 vs 12 symbols, again controlling for non-numerical factors.

Many studies have shown that when given a choice, shoaling fish prefer to join the largest of two shoals. It has been argued that several aspects of such choice reflect an ability by fish to distinguish between numerical quantities.

Social Learning

Fish can learn how to perform a behavior simply by watching other individuals in action. This is variously called observational learning, cultural transmission, or social learning. For example, fish can learn a particular route after following an experienced leader a few times. One study trained guppies to swim through a hole marked in red while ignoring another one marked in green in order to get food on the other side of a partition; when these experienced fish ("demonstrators") were joined by a naive one (an "observer"), the observer followed the demonstrators through the red hole, and kept the habit once the demonstrators were removed, even when the green hole now allowed food access. In the wild, juvenile French grunt follow traditional migration routes, up to 1 km long, between their daytime resting sites and their nighttime foraging areas on coral reefs; if groups of 10-20 individuals are marked and then transplanted to new populations, they follow the residents along what is for them – the transplants – a new migration route, and if the residents are then removed two days later, the transplanted grunts continue to use the new route, as well as the resting and foraging sites at both ends.

Through cultural transmission, fishes could also learn where good food spots are. Ninespine stickleback, when given a choice between two food patches they have watched for a while, prefer the patch over which more fish have been seen foraging, or over which fish were seen feeding more intensively. Similarly, in a field experiment where Trinidadian guppies were given a choice between two distinctly marked feeders in their home rivers, the subjects chose the feeder where other guppies were already present, and in subsequent tests when both feeders were deserted, the subjects remembered the previously popular feeder and chose it.

Through social learning, fishes might learn not only where to get food, but also what to get and how to get it. Hatchery-raised salmon can be taught to quickly accept novel, live prey items similar to those they will encounter once they will be released in the wild, simply by watching an experienced salmon take such prey. The same is true of young perch. In the laboratory, juvenile European seabass can learn to push a lever in order to obtain food just by watching experienced individuals use the lever.

Fishes can also learn from others the identity of predatory species. Fathead minnows, for example, can learn the smell of a predatory pike just by being simultaneously exposed to that smell and the sight of experienced minnows reacting with fear, and brook stickleback can learn the visual identity of a predator by watching the fright reaction of experienced fathead minnows. Fish can also learn to recognize the odor of dangerous sites when they are simultaneously exposed to it and to other fish that suddenly show a fright reaction. Hatchery-raised salmon can learn the smell of a predator by being simultaneously exposed to it and to the alarm substance released by injured salmon.

Latent Learning

Latent learning is a form of learning that is not immediately expressed in an overt response; it occurs without any obvious reinforcement of the behaviour or associations that are learned. One example in fish comes from research with male three spot gouramis (*Trichopodus trichopterus*). This species quickly form dominance hierarchies. To appease dominants, subordinates adopt a typical body posture angled at 15-60° to the horizontal, all fins folded and pale body colors. Individuals trained to associate a light-stimulus with the imminent arrival of food exhibit this associative learning by approaching the surface where the food is normally dropped immediately the light-stimulus is presented. However, if a subordinate is placed in a tank with a dominant individual and the light-stimulus is presented, the subordinate immediately assumes the submissive posture rather than approaching the surface. The subordinate has predicted that going to the surface to get food would place it in competition with the dominant, and to avoid potential aggression, it immediately attempts to appease the dominant.

Cleaner Fish

The bluestreak cleaner wrasse (*Labroides dimidiatus*) performs a service for "client" fishes (belonging to other species) by removing and eating their ectoparasites. Clients can invite a cleaning session by adopting a typical posture or simply by remaining immobile near a wrasse's cleaning station. They can even form queues while doing so. But cleaning sessions do not always end up well, because wrasses (or wrasse-mimicking parasitic sabre-toothed blennies) may cheat and eat the nutritious body mucus of their clients, rather than just the ectoparasites, something that makes the client jolt and sometimes flee. This system has been the subject of extensive observations which have suggested cognitive abilities on the part of the cleaner wrasses and their clients. For example, clients refrain from soliciting a cleaning session if they have witnessed the cleaning session of the previous client ending badly. Cleaners give the impression of trying to maintain a good reputation, because they cheat less when they see a big audience (a long queue of clients) watching. Cleaners sometimes work as male-female teams, and when the smaller female cheats and bites the client, the larger male chases her off, as if to punish her for having tarnished their reputation.

Bluestreak cleaner wrasse (bottom) with a client fish

Play

Play behaviour is often considered a correlate of intelligence. One possible example in fish is provided by the electrolocating Peters' elephantnose fish (mentioned above as having one of the largest brain-to-body weight ratios of all known vertebrates). One captive individual was observed carrying a small ball of aluminum foil (a good conductor of electricity) to the outflow tube of the aquarium filter, letting the current push the ball away before chasing after it and repeating the behaviour.

Food Stocking

Food stocking can be viewed potentially as an animal planning for the future. One example of short-term stocking involves climbing perch (*Anabas testudineus*). Individuals were kept singly in aquaria and fed with pellets dropped at the surface. When the pellets were dropped one after the other at 1-s intervals, the fish took them as they reached the surface and stocked them inside the mouth. On average, the fish placed 7 pellets in their mouth before moving away to consume them. When starved for 24-h before the feeding test, they doubled the number of pellets stocked (14 on average); the underside of their heads bulged under the load. The behaviour may be an indication that competition for food is normally severe in this species and that any adaptation to secure food would be beneficial.

References

- Schwab, Ivan R (2002). "More than just cool shades". British Journal of Ophthalmology. 86: 1075. doi:10.1136/bjo.86.10.1075

- Prosser, C. Ladd (1991). Comparative Animal Physiology, Environmental and Metabolic Animal Physiology (4th ed.). Hoboken, NJ: Wiley-Liss. pp. 1–12. ISBN 0-471-85767-X

- "Modifications of the Digestive Tract for Holding Air in Loricariid and Scoloplacid Catfishes" (PDF). Copeia (3): 663–675. 1998. doi:10.2307/1447796. Retrieved 25 June 2009

- "The Origin of the Larva and Metamorphosis in Amphibia". The American Naturalist. Essex Institute. 91: 287. 1957. JSTOR 2458911. doi:10.1086/281990

- Meyer CG, Holland KN, Papastamatiou YP (2005). "Sharks can detect changes in the geomagnetic field". Journal of the Royal Society, Interface. 2 (2): 129–30. PMC 1578252. PMID 16849172. doi:10.1098/rsif.2004.0021

- Hoar WS and Randall DJ (1984) Fish Physiology: Gills: Part A – Anatomy, gas transfer and acid-base regulation Academic Press. ISBN 9780080585314

- Fowler, Sarah; Cavanagh, Rachel. "Sharks, Rays, and Chimaeras: The Status of Chondrichthyan Fishes" (PDF). The World Conservation Union. Retrieved 3 May 2017

- Reddon, AR; O'Connor, CM; Marsh-Rollo, SE; Balcshine, S (2012). "Effects of isotocin on social responses in a cooperatively breeding fish". Animal Behaviour. 84 (4): 753–760. doi:10.1016/j.anbehav.2012.07.021

- Märss, T. (2006). "Exoskeletal ultrasculpture of early vertebrates". Journal of Vertebrate Paleontology. 26 (2): 235–252. doi:10.1671/0272-4634(2006)26[235:EUOEV]2.0.CO;2

- Janvier, Philippe (1998). "Early vertebrates and their extant relatives". Early Vertebrates. Oxford University Press. pp. 123–127. ISBN 0-19-854047-7

- "Rose, J.D. 2003. A Critique of the paper: "Do fish have nociceptors: Evidence for the evolution of a vertebrate sensory system"" (PDF). Retrieved 21 May 2011

- "The Origin of the Larva and Metamorphosis in Amphibia". The American Naturalist. Essex Institute. 91: 287. 1957. JSTOR 2458911. doi:10.1086/281990

- Reimchen, T E; Temple, N F (2004). "Hydrodynamic and phylogenetic aspects of the adipose fin in fishes". Canadian Journal of Zoology. 82 (6): 910–916. doi:10.1139/Z04-069

- Kapoor BG and Khanna B (2004) Ichthyology Handbook pp. 497–498, Springer Science & Business Media. ISBN 9783540428541

- Dement J Species Spotlight: Atlantic Sailfish (Istiophorus albicans) Archived December 17, 2010, at the Wayback Machine. littoralsociety.org. Retrieved 1 April 2012

- Sharpe, P. T. (2001). "Fish scale development: Hair today, teeth and scales yesterday?". Current Biology. 11 (18): R751–R752. PMID 11566120. doi:10.1016/S0960-9822(01)00438-9

- Nauen, JC; Lauder, GV (2000). "Locomotion in scombrid fishes: morphology and kinematics of the finlets of the Chub mackerel Scomber japonicus" (PDF). Journal of Experimental Biology. 203: 2247–59

- Gould,Stephen Jay (1993) "Bent Out of Shape" in Eight Little Piggies: Reflections in Natural History. Norton, 179–94. ISBN 9780393311396

- Purple Flying Gurnard, Dactyloptena orientalis (Cuvier, 1829) Australian Museum. Updated: 15 September 2012. Retrieved: 2 November 2012

- Coates, M. I. (2003). "The Evolution of Paired Fins". Theory in Biosciences. 122 (2–3): 266–87. doi:10.1078/1431-7613-00087

Fish Farming: An Essential Aspect

Fish farming is the raising of fishes for commercial purposes. The commonly farmed fishes are Atlantic salmon, sea bass, turbot and trout. This chapter discusses the methods of fish farming in a critical manner providing key analysis to the subject matter.

Fish Farming

Salmon farming in the sea (mariculture) at Loch Ainort, Isle of Skye

Fish farming or pisciculture involves raising fish commercially in tanks or enclosures, usually for food. It is the principal form of aquaculture, while other methods may fall under mariculture. A facility that releases juvenile fish into the wild for recreational fishing or to supplement a species' natural numbers is generally referred to as a fish hatchery. Worldwide, the most important fish species used in fish farming are carp, tilapia, salmon, and catfish.

Demand is increasing for fish and fish protein, which has resulted in widespread overfishing in wild fisheries. China provides 62% of the world's farmed fish. As of 2016, more than 50% of seafood was produced by aquaculture.

Farming carnivorous fish, such as salmon, does not always reduce pressure on wild fisheries, since carnivorous farmed fish are usually fed fishmeal and fish oil extracted from wild forage fish. The 2008 global returns for fish farming recorded by the FAO totaled 33.8 million tonnes worth about $US 60 billion.

Major Species

Top 15 cultured fish species by weight in millions of tonnes, according to FAO statistics for 2013			
Species	Environment	Tonnage (millions)	Value (US$, billion)
Grass carp	freshwater	5.23	6.69
Silver carp	freshwater	4.59	6.13
Common carp	freshwater	3.76	5.19
Nile tilapia	freshwater	3.26	5.39
Bighead carp	freshwater	2.90	3.72
Catla (Indian carp)	freshwater	2.76	5.49
Crucian carp	freshwater	2.45	2.67
Atlantic salmon	marine	2.07	10.10
Roho labeo	freshwater	1.57	2.54
Milkfish	freshwater	0.94	1.71
Rainbow trout	freshwater, brackish, marine	0.88	3.80
Wuchang bream	freshwater	0.71	1.16
Black carp	freshwater	0.50	1.15
Northern snakehead	freshwater	0.48	0.59
Amur catfish	freshwater	0.41	0.55

Categories

Aquaculture makes use of local photosynthetic production (extensive) or fish that are fed with external food supply (intensive).

Extensive Aquaculture

Aqua-Boy, a Norwegian live fish carrier used to serve the Marine Harvest fish farms on the west coast of Scotland

Growth is limited by available food, commonly zooplankton feeding on pelagic algae or benthic animals, such as crustaceans and mollusks. Tilapia filter feed directly on phytoplankton, which makes higher production possible. Photosynthetic production can be increased by fertilizing pond water with artificial fertilizer mixtures, such as potash, phosphorus, nitrogen, and microelements.

Another issue is the risk of algal blooms. When temperatures, nutrient supply, and available sunlight are optimal for algal growth, algae multiply at an exponential rate, eventually exhausting nutrients and causing a subsequent die-off. The decaying algal biomass depletes the oxygen in the pond water because it blocks out the sun and pollutes it with organic and inorganic solutes (such as ammonium ions), which can (and frequently do) lead to massive loss of fish.

An alternate option is to use a wetland system such as that of Veta la Palma in Spain.

To tap all available food sources in the pond, the aquaculturist chooses fish species which occupy different places in the pond ecosystem, e.g., a filter algae feeder such as tilapia, a benthic feeder such as carp or [catfish and a zooplankton feeder (various carps) or submerged weeds feeder such as grass carp.

Despite these limitations, significant fish farming industries use these methods. In the Czech Republic, thousands of natural and seminatural ponds are harvested each year for trout and carp. The large ponds around Trebon built from around 1650 are still in use.

Intensive Aquaculture

Optimal water parameters for cold- and warm-water fish in intensive aquaculture	
Acidity	pH 6-9
Arsenic	<440 µg/l
Alkalinity	>20 mg/l (as $CaCO_3$)
Aluminum	<0.075 mg/l
Ammonia (non-ionized)	<0.02mg/l
Cadmium	<0.0005 mg/l in soft water; < 0.005 mg/L in hard water
Calcium	>5 mg/l
Carbon dioxide	<5–10 mg/l
Chloride	>4.0 mg/l
Chlorine	<0.003 mg/l
Copper	<0.0006 mg/l in soft water; < 0.03 mg/l in hard water
Gas supersaturation	<100% total gas pressure (103% for salmonid eggs/fry) (102% for lake trout)
Hydrogen sulfide	<0.003 mg/l
Iron	<0.1 mg/l
Lead	<0.02 mg/l
Mercury	<0.0002 mg/l
Nitrate	<1.0 mg/l
Nitrite	<0.1 mg/l
Oxygen	6 mg/l for coldwater fish 4 mg/l for warmwater fish
Selenium	<0.01 mg/l
Total dissolved solids	<200 mg/l
Total suspended solids	<80 NTU over ambient levels
Zinc	<0.005 mg/l

In these kinds of systems fish production per unit of surface can be increased at will, as long as sufficient oxygen, fresh water and food are provided. Because of the requirement of sufficient fresh water, a massive water purification system must be integrated in the fish farm. One way to achieve this is to combine hydroponic horticulture and water treatment, see below. The exception to this rule are cages which are placed in a river or sea, which supplements the fish crop with sufficient oxygenated water. Some environmentalists object to this practice.

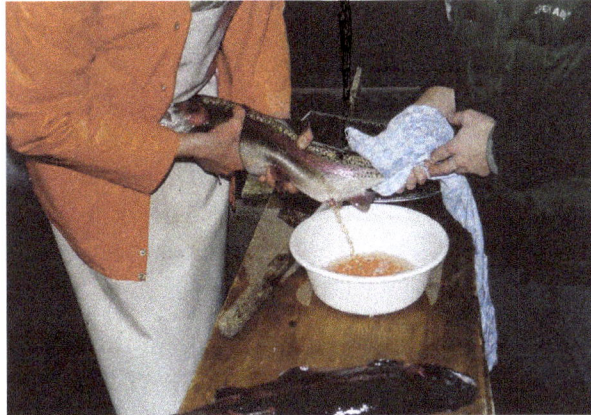
Expressing eggs from a female rainbow trout

The cost of inputs per unit of fish weight is higher than in extensive farming, especially because of the high cost of fish feed, which must contain a much higher level of protein (up to 60%) than cattle feed and a balanced amino acid composition, as well. However, these higher protein level requirements are a consequence of the higher feed efficiency of aquatic animals (higher feed conversion ratio [FCR], that is, kg of feed per kg of animal produced). Fish such as salmon have an FCR around 1.1 kg of feed per kg of salmon whereas chickens are in the 2.5 kg of feed per kg of chicken range. Fish do not use energy to keep warm, eliminating some carbohydrates and fats in the diet, required to provide this energy. This may be offset, though, by the lower land costs and the higher production which can be obtained due to the high level of input control.

Aeration of the water is essential, as fish need a sufficient oxygen level for growth. This is achieved by bubbling, cascade flow, or aqueous oxygen. *Clarias* spp. can breathe atmospheric air and can tolerate much higher levels of pollutants than trout or salmon, which makes aeration and water purification less necessary and makes *Clarias* species especially suited for intensive fish production. In some *Clarias* farms, about 10% of the water volume can consist of fish biomass.

The risk of infections by parasites such as fish lice, fungi (*Saprolegnia* spp.), intestinal worms (such as nematodes or trematodes), bacteria (e.g., *Yersinia* spp., *Pseudomonas* spp.), and protozoa (such as dinoflagellates) is similar to animal husbandry, especially at high population densities. However, animal husbandry is a larger and more technologically mature area of human agriculture and better solutions to pathogen problem exist. Intensive aquaculture does have to provide adequate water quality (oxygen, ammonia, nitrite, etc.) levels to minimize stress, which makes the pathogen problem more difficult. This means intensive aquaculture requires tight monitoring and a high level of expertise of the fish farmer.

Very-high-intensity recycle aquaculture systems (RAS), where all the production parameters are controlled, are being used for high-value species. By recycling it, very little water is used per unit of production. However, the process does have high capital and operating costs. The higher cost structures mean that RAS is only economical for high-value products such as broodstock for egg production, fingerlings for net pen aquaculture operations, sturgeon production, research animals, and some special niche markets like live fish.

Controlling roes manually

Raising ornamental coldwater fish (goldfish or koi), although theoretically much more profitable due to the higher income per weight of fish produced, has never been successfully carried out until recently. The increased incidences of dangerous viral diseases of koi carp, together with the high value of the fish, has led to initiatives in closed-system koi breeding and growing in a number of countries. Today, a few commercially successful intensive koi growing facilities exist in the UK, Germany, and Israel.

Some producers have adapted their intensive systems in an effort to provide consumers with fish that do not carry dormant forms of viruses and diseases.

In 2016, juvenile Nile tilapia were given a food containing dried *Schizochytrium* in place of fish oil. When compared to a control group raised on regular food, they exhibited higher weight gain and better food-to-growth conversion, plus their flesh was higher in healthy omega-3 fatty acids.

Fish Farms

Within intensive and extensive aquaculture methods, numerous specific types of fish farms are used; each has benefits and applications unique to its design.

Cage System

Giant gourami is often raised in cages in central Thailand.

Fish cages are placed in lakes, bayous, ponds, rivers, or oceans to contain and protect fish until they can be harvested. The method is also called "off-shore cultivation" when the cages are placed in the sea. They can be constructed of a wide variety of components. Fish are stocked in cages, artificially fed, and harvested when they reach market size. A few advantages of fish farming with cages are that many types of waters can be used (rivers, lakes, filled quarries, etc.), many types of fish can be raised, and fish farming can co-exist with sport fishing and other water uses. Cage farming of fishes in open seas is also gaining popularity. Concerns of disease, poaching, poor water quality, etc., lead some to believe that in general, pond systems are easier to manage and simpler to start. Also, past occurrences of cage-failures leading to escapes, have raised concern regarding the culture of non-native fish species in dam or open-water cages. Though the cage-industry has made numerous technological advances in cage construction in recent years, storms always make the concern for escapes valid.

Recently, copper alloys have become important netting materials in aquaculture. Copper alloys are antimicrobial, that is, they destroy bacteria, viruses, fungi, algae, and other microbes. In the marine environment, the antimicrobial/algaecidal properties of copper alloys prevent biofouling, which can briefly be described as the undesirable accumulation, adhesion, and growth of microorganisms, plants, algae, tube worms, barnacles, mollusks, and other organisms.

The resistance of organism growth on copper alloy nets also provides a cleaner and healthier environment for farmed fish to grow and thrive. Traditional netting involves regular and labor-intensive cleaning. In addition to its antifouling benefits, copper netting has strong structural and corrosion-resistant properties in marine environments.

Copper-zinc brass alloys are deployed in commercial-scale aquaculture operations in Asia, South America, and the USA (Hawaii). Extensive research, including demonstrations and trials, are being implemented on two other copper alloys: copper-nickel and copper-silicon. Each of these alloy types has an inherent ability to reduce biofouling, cage waste, disease, and the need for antibiotics, while simultaneously maintaining water circulation and oxygen requirements. Other types of copper alloys are also being considered for research and development in aquaculture operations.

Irrigation Ditch or Pond Systems

These fish-farming ponds were created as a cooperative project in a rural village.

These use irrigation ditches or farm ponds to raise fish. The basic requirement is to have a ditch or pond that retains water, possibly with an above-ground irrigation system (many irrigation systems use buried pipes with headers.)

Using this method, water allotments can be stored in ponds or ditches, usually lined with bentonite clay. In small systems, the fish are often fed commercial fish food, and their waste products can help fertilize the fields. In larger ponds, the pond grows water plants and algae as fish food. Some of the most successful ponds grow introduced strains of plants, as well as introduced strains of fish.

Control of water quality is crucial. Fertilizing, clarifying, and pH control of the water can increase yields substantially, as long as eutrophication is prevented and oxygen levels stay high. Yields can be low if the fish grow ill from electrolyte stress.

Composite Fish Culture

The composite fish culture system is a technology developed in India by the Indian Council of Agricultural Research in the 1970s. In this system, of both local and imported fish, a combination of five or six fish species is used in a single fish pond. These species are selected so that they do not compete for food among them by having different types of food habitats. As a result, the food available in all the parts of the pond is used. Fish used in this system include catla and silver carp which are surface feeders, rohu, a column feeder, and mrigal and common carp, which are bottom feeders. Other fish also feed on the excreta of the common carp, and this helps contribute to the efficiency of the system which in optimal conditions produces 3000–6000 kg of fish per hectare per year.

One problem with such composite fish culture is that many of these fish breed only during monsoon. Even if fish are collected from the wild, they can be mixed with other species, as well. So, a major problem in fish farming is the lack of availability of good-quality stock. To overcome this problem, ways have now been worked out to breed these fish in ponds using hormonal stimulation. This has ensured the supply of pure fish stock in desired quantities.

Integrated Recycling Systems

Aerators in a fish farm (Ararat plain, Armenia)

One of the largest problems with freshwater pisciculture is that it can use a million gallons of water per acre (about 1 m³ of water per m²) each year. Extended water purification systems allow for the reuse (recycling) of local water.

The largest-scale pure fish farms use a system derived (admittedly much refined) from the New Alchemy Institute in the 1970s. Basically, large plastic fish tanks are placed in a greenhouse. A hydroponic bed is placed near, above or between them. When tilapia are raised in the tanks, they are able to eat algae, which naturally grow in the tanks when the tanks are properly fertilized.

The tank water is slowly circulated to the hydroponic beds, where the tilapia waste feeds commercial plant crops. Carefully cultured microorganisms in the hydroponic bed convert ammonia to nitrates, and the plants are fertilized by the nitrates and phosphates. Other wastes are strained out by the hydroponic media, which double as an aerated pebble-bed filter.

This system, properly tuned, produces more edible protein per unit area than any other. A wide variety of plants can grow well in the hydroponic beds. Most growers concentrate on herbs (e.g. parsley and basil), which command premium prices in small quantities all year long. The most common customers are restaurant wholesalers.

Since the system lives in a greenhouse, it adapts to almost all temperate climates, and may also adapt to tropical climates. The main environmental impact is discharge of water that must be salted to maintain the fishes' electrolyte balance. Current growers use a variety of proprietary tricks to keep fish healthy, reducing their expenses for salt and wastewater discharge permits. Some veterinary authorities speculate that ultraviolet ozone disinfectant systems (widely used for ornamental fish) may play a prominent part in keeping the tilapia healthy with recirculated water.

A number of large, well-capitalized ventures in this area have failed. Managing both the biology and markets is complicated. One future development is the combination of integrated recycling systems with urban farming as tried in Sweden by the Greenfish Initiative.

Classic Fry Farming

This is also called a "flow through system" Trout and other sport fish are often raised from eggs to fry or fingerlings and then trucked to streams and released. Normally, the fry are raised in long, shallow, concrete tanks, fed with fresh stream water. The fry receive commercial fish food in pellets. While not as efficient as the New Alchemists' method, it is also far simpler, and has been used for many years to stock streams with sport fish. European eel (*Anguilla anguilla*) aquaculturalists procure a limited supply of glass eels, juvenile stages of the European eel which swim north from the Sargasso Sea breeding grounds, for their farms. The European eel is threatened with extinction because of the excessive catch of glass eels by Spanish fishermen and overfishing of adult eels in, e.g., the Dutch IJsselmeer, Netherlands. As of 2005, no one has managed to breed the European eel in captivity.

Issues

The issue of feeds in fish farming has been a controversial one. Many cultured fishes (tilapia, carp, catfish, many others) require no meat or fish products in their diets. Top-level carnivores (most salmon species) depend on fish feed of which a portion is usually derived from wild-caught fish (anchovies, menhaden, etc.). Vegetable-derived proteins have successfully replaced fish meal in feeds for carnivorous fishes, but vegetable-derived oils have not successfully been incorporated into the diets of carnivores. Research is underway to try to change this, such that even salmon and other carnivores could be successfully fed with vegetable products. The F3 Challenge (Fish-Free Feed Challenge), as explained by a report from *Wired* in February 2017, "is a race to sell 100,000 metric tons of fish food, without the fish. Earlier this month, start-ups from places like Pakistan, China, and Belgium joined their American competition at the Google headquarters in Mountain View, CA, showing off feed made from seaweed extracts, yeast, and algae grown in bioreactors." However, it will be some years before these efforts have industry-wide effect, if they succeed.

Secondly, farmed fish are kept in concentrations never seen in the wild (e.g. 50,000 fish in a 2-acre (8,100 m²) area.). However, fish tend also to be animals that aggregate into large schools at high density. Most successful aquaculture species are schooling species, which do not have social problems at high density. Aquaculturists feel that operating a rearing system above its design capacity or above the social density limit of the fish will result in decreased growth rate and increased feed conversion ratio (kg dry feed/kg of fish produced), which results in increased cost and risk of health problems along with a decrease in profits. Stressing the animals is not desirable, but the concept of and measurement of stress must be viewed from the perspective of the animal using the scientific method.

Sea lice, particularly *Lepeophtheirus salmonis* and various *Caligus* species, including *C. clemensi* and *C. rogercresseyi*, can cause deadly infestations of both farm-grown and wild salmon. Sea lice are ectoparasites which feed on mucus, blood, and skin, and migrate and latch onto the skin of wild salmon during free-swimming, planktonic *nauplii* and *copepodid* larval stages, which can persist for several days. Large numbers of highly populated, open-net salmon farms can create exceptionally large concentrations of sea lice; when exposed in river estuaries containing large numbers of open-net farms, many young wild salmon are infected, and do not survive as a result. Adult salmon may survive otherwise critical numbers of sea lice, but small, thin-skinned juvenile salmon migrating to sea are highly vulnerable. On the Pacific coast of Canada, the louse-induced mortality of pink salmon in some regions is commonly over 80%.

A 2008 meta-analysis of available data shows that salmon farming reduces the survival of associated wild salmon populations. This relationship has been shown to hold for Atlantic, steelhead, pink, chum, and coho salmon. The decrease in survival or abundance often exceeds 50%.

Diseases and parasites are the most commonly cited reasons for such decreases. Some species of sea lice have been noted to target farmed coho and Atlantic salmon. Such parasites have been shown to have an effect on nearby wild fish. One place that has garnered international media attention is British Columbia's Broughton Archipelago. There, juvenile wild salmon must "run a gauntlet" of large fish farms located off-shore near river outlets before making their way to sea. The farms allegedly cause such severe sea lice infestations that one study predicted in 2007 a 99% collapse in the wild salmon population by 2011. This claim, however, has been criticized by numerous scientists who question the correlation between increased fish farming and increases in sea lice infestation among wild salmon.

Because of parasite problems, some aquaculture operators frequently use strong antibiotic drugs to keep the fish alive, but many fish still die prematurely at rates up to 30%. In some cases, these drugs have entered the environment. Additionally, the residual presence of these drugs in human food products has become controversial. Use of antibiotics in food production is thought to increase the prevalence of antibiotic resistance in human diseases. At some facilities, the use of antibiotic drugs in aquaculture has decreased considerably due to vaccinations and other techniques. However, most fish-farming operations still use antibiotics, many of which escape into the surrounding environment.

The lice and pathogen problems of the 1990s facilitated the development of current treatment methods for sea lice and pathogens, which reduced the stress from parasite/pathogen problems. However, being in an ocean environment, the transfer of disease organisms from the wild fish to the aquaculture fish is an ever-present risk.

The large number of fish kept long-term in a single location contributes to habitat destruction of the nearby areas. The high concentrations of fish produce a significant amount of condensed faeces, often contaminated with drugs, which again affects local waterways. However, if the farm is correctly placed in an area with a strong current, the 'pollutants' are flushed out of the area fairly quickly. Not only does this help with the pollution problem, but water with a stronger current also aids in overall fish growth. Concern remains that resultant bacterial growth strips the water of oxygen, reducing or killing off the local marine life. Once an area has been so contaminated, the fish farms are moved to new, uncontaminated areas. This practice has angered nearby fishermen.

Other potential problems faced by aquaculturists are the obtaining of various permits and water-use rights, profitability, concerns about invasive species and genetic engineering depending on what species are involved, and interaction with the United Nations Convention on the Law of the Sea.

In regards to genetically modified, farmed salmon, concern has been raised over their proven reproductive advantage and how it could potentially decimate local fish populations, if released into the wild. Biologist Rick Howard did a controlled laboratory study where wild fish and GMO fish were allowed to breed. The GMO fish crowded out the wild fish in spawning beds, but the offspring were less likely to survive. The colorant used to make pen-raised salmon appear rosy like the wild fish has been linked with retinal problems in humans.

Labeling

In 2005, Alaska passed legislation requiring that any genetically altered fish sold in the state be labeled. In 2006, a *Consumer Reports* investigation revealed that farm-raised salmon is frequently sold as wild.

In 2008, the US National Organic Standards Board allowed farmed fish to be labeled as organic provided less than 25% of their feed came from wild fish. This decision was criticized by the advocacy group Food & Water Watch as "bending the rules" about organic labeling. In the European Union, fish labeling as to species, method of production and origin, has been required since 2002.

Concerns continue over the labeling of salmon as farmed or wild-caught, as well as about the humane treatment of farmed fish. The Marine Stewardship Council has established an Eco label to distinguish between farmed and wild-caught salmon, while the RSPCA has established the Freedom Food label to indicate humane treatment of farmed salmon, as well as other food products.

Indoor Fish Farming

An alternative to outdoor open ocean cage aquaculture, is through the use of a recirculating aquaculture system (RAS). A RAS is a series of culture tanks and filters where water is continuously recycled and monitored to keep optimal conditions year round. To prevent the deterioration of water quality, the water is treated mechanically through the removal of particulate matter and biologically through the conversion of harmful accumulated chemicals into nontoxic ones.

Other treatments such as ultraviolet sterilization, ozonation, and oxygen injection are also used to maintain optimal water quality. Through this system, many of the environmental drawbacks of aquaculture are minimized including escaped fish, water usage, and the introduction of pollutants. The practices also increased feed-use efficiency growth by providing optimum water quality.

One of the drawbacks to recirculating aquaculture systems is the need for periodic water exchanges. However, the rate of water exchange can be reduced through aquaponics, such as the incorporation of hydroponically grown plants and denitrification. Both methods reduce the amount of nitrate in the water, and can potentially eliminate the need for water exchanges, closing the aquaculture system from the environment. The amount of interaction between the aquaculture system and the environment can be measured through the cumulative feed burden (CFB kg/M3), which measures the amount of feed that goes into the RAS relative to the amount of water and waste discharged.

From 2011, a team from the University of Waterloo led by Tahbit Chowdhury and Gordon Graff examined vertical RAS aquaculture designs aimed at producing protein-rich fish species. However, because of its high capital and operating costs, RAS has generally been restricted to practices such as broodstock maturation, larval rearing, fingerling production, research animal production, specific pathogen-free animal production, and caviar and ornamental fish production. As such, research and design work by Chowdhury and Graff remains difficult to implement. Although the use of RAS for other species is considered by many aquaculturalists to be currently impractical, some limited successful implementation of RAS has occurred with high-value product such as barramundi, sturgeon, and live tilapia in the US eels and catfish in the Netherlands, trout in Denmark and salmon is planned in Scotland and Canada.

Slaughter Methods

Tanks saturated with carbon dioxide have been used to make fish unconscious. Their gills are then cut with a knife so that the fish bleed out before they are further processed. This is no longer considered a humane method of slaughter. Methods that induce much less physiological stress are electrical or percussive stunning and this has led to the phasing out of the carbon dioxide slaughter method in Europe.

Inhumane Methods

According to T. Håstein of the National Veterinary Institute, "Different methods for slaughter of fish are in place and it is no doubt that many of them may be considered as appalling from an animal welfare point of view." A 2004 report by the EFSA Scientific Panel on Animal Health and Welfare explained: "Many existing commercial killing methods expose fish to substantial suffering over a prolonged period of time. For some species, existing methods, whilst capable of killing fish humanely, are not doing so because operators don't have the knowledge to evaluate them." Following are some of the less humane ways of killing fish.

- Air asphyxiation amounts to suffocation in the open air. The process can take upwards of 15 minutes to induce death, although unconsciousness typically sets in sooner.

- Ice baths or chilling of farmed fish on ice or submerged in near-freezing water is used to dampen muscle movements by the fish and to delay the onset of post-death decay. However, it does not necessarily reduce sensibility to pain; indeed, the chilling process has been shown to elevate cortisol. In addition, reduced body temperature extends the time before fish lose consciousness.

- CO_2 narcosis

- Exsanguination without stunning is a process in which fish are taken up from water, held still, and cut so as to cause bleeding. According to references in Yue, this can leave fish writhing for an average of four minutes, and some catfish still responded to noxious stimuli after more than 15 minutes.

- Immersion in salt followed by gutting or other processing such as smoking is applied to eel.

More Humane Methods

Proper stunning renders the fish unconscious immediately and for a sufficient period of time such that the fish is killed in the slaughter process (e.g. through exsanguination) without regaining consciousness.

- Percussive stunning involves rendering the fish unconscious with a blow on the head.

- Electric stunning can be humane when a proper current is made to flow through the fish brain for a sufficient period of time. Electric stunning can be applied after the fish has been taken out of the water (dry stunning) or while the fish is still in the water. The latter generally requires a much higher current and may lead to operator safety issues. An advantage could be that in-water stunning allows fish to be rendered unconscious without stressful handling or displacement. However, improper stunning may not induce insensibility long enough to prevent the fish from enduring exsanguination while conscious. Whether the optimal stunning parameters that researchers have determined in studies are used by the industry in practice is unknown.

Aquaculture of Catfish

Loading U.S. farm-raised catfish

Catfish are easy to farm in warm climates, leading to inexpensive and safe food at local grocers. Catfish raised in inland tanks or channels are considered safe for the environment, since their waste and disease should be contained and not spread to the wild.

Asia

In Asia, many catfish species are important as food. Several walking catfish (Clariidae) and shark catfish (Pangasiidae) species are heavily cultured in Africa and Asia. Exports of one particular shark catfish species from Vietnam, *Pangasius bocourti*, has met with pressures from the U.S. catfish industry. In 2003, The United States Congress passed a law preventing the imported fish from being labeled as catfish. As a result, the Vietnamese exporters of this fish now label their products sold in the U.S. as "basa fish".

United States

Ictalurids are cultivated in North America (especially in the Deep South, with Mississippi being the largest domestic catfish producer). Channel catfish (*Ictalurus punctatus*) supports a $450 million/ yr aquaculture industry. The US farm-raised catfish industry began in the early 1960s in Kansas, Oklahoma and Arkansas. Channel catfish quickly became the major catfish grown, as it was hardy and easily spawned in earthen ponds. By the late 60s, the industry moved into the Mississippi Delta as farmers struggled with sagging profits in cotton, rice and soybeans, especially on those farm areas where soils had a very high clay content.

The Mississippi Deltaic Plain includes two active pro-grading deltas: the modern bird-foot [Balize] delta, commonly referred to as the Mississippi Delta, and the Atchafalaya delta. In addition, there are degrading deltaic systems, such as the Lafourche and the St. Bernard [ref: World Delta Data Base, Hart and Coleman]. These deltas became the industry home for the catfish industry, as they had the soils, climate and shallow aquifers to provide water for the earthen ponds that grow 360-380 million pounds (160,000 to 170,000 tons) of catfish annually. Catfish are fed a grain-based diet that includes soybean meal. Fish are fed daily through the summer at rates of 1-6% of body weight with the pelleted floating feed. Catfish need about two pounds of feed to produce one pound of live weight. Mississippi is home to 100,000 acres (400 km²) of catfish ponds, the largest of any state. Other states important in growing catfish include Alabama, Arkansas and Louisiana.

Aquarium

There is a large and growing ornamental fish trade, with hundreds of species of catfish, such as *Corydoras* and armored suckermouth catfish (often called plecos), being a popular component of many aquaria. Other catfish commonly found in the aquarium trade are banjo catfish, talking catfish, and long-whiskered catfish.

Aquaculture of Cobia

Cobia, a warm water fish, is one of the more suitable candidates for offshore aquaculture. Cobia are large pelagic fish, up to 2 metres (78 inches) long and 68 kilograms (150 pounds) in weight. They are solitary fish except when spawning, found in warm-temperate to tropical waters.

A female broodstock cobia weighing about 8 kilograms prior to transport to broodstock holding tanks

Their rapid growth rate in aquaculture, as well as the high quality of their flesh, makes cobia potentially one of the more important potential marine fish for aquaculture production. Currently, cobia are cultured in nurseries and grow-out offshore cages in many parts of Asia and off the coast of the United States, Mexico and Panama. In Taiwan cobia weighing 100–600 grams are cultured for 1–1.5 years to reach the 6–8 kilograms needed for export to Japan. Currently, around 80% of marine cages in Taiwan are devoted to cobia culture. In 2004, the FAO reported that 80.6% of the world's cobia production was by China and Taiwan. After China and Taiwan, Vietnam is the third largest producer of farmed cobia in the world where production was estimated at 1500 tonnes in 2008. The possibility is also being examined of growing hatchery reared cobia in offshore cages around Puerto Rico and the Bahamas.

Greater depths, stronger currents, and distance from shore all act to reduce the environmental impacts often associated with fin fish aquaculture. Offshore cage systems could become some of the most environmentally sustainable methods for commercial marine fish aquaculture. However, some problems still exist in cobia culture that needs to be addressed and solved for increasing production. These include high mortality rates due to stress during transport from nursery tanks or inshore cages out to grow-out cages. Also, diseases in the nursery stage and the grow-out culture can result in low survival rates and a poor harvest.

Production

Cobia fingerlings in aquaculture

Wild cobia broodstock are captured by professional fishermen. The fish are transferred into on-board-tanks on a transport vessel for transport to hatchery facilities. They are anesthetized with clove oil if necessary to reduce stress during transportation. They are also treated for ectoparasites on their gills and skin that could proliferate later after transfer to maturation tanks.

Broodstock are reared in controlled ponds or tanks. These tanks are often stocked with cleaner fish, *Gobiosoma oceanops*, as a biological control for any remaining ectoparasites. The broodstock diet includes sardines, squid and formulated feeds, as well as vitamin and mineral supplements. The water temperature is used to control spawning.

The eggs are collected with a surface skimmer using mesh screen bags. The eggs are transferred to incubation tanks where they are disinfected for an hour with 100 ppm formalin.

Phytoplankton concentrations are maintained, and enriched *Artemia* nauplii and rotifers are fed to the cobia larvae for 3–7 days after they hatch. The larvae require rotifers for at least four days after hatching. The presence of enriched live prey in conjunction with live algae in rearing tanks has been shown to improve the way larvae grow and survive in recirculating systems.

Optimal rearing densities are required when rearing larvae. Even though water quality and food can be controlled, it has been shown that high rearing densities may still affect growth and survival of the larvae through responses related to crowding. In addition, juveniles exposed to varying salinities exhibited sustained growth and improved health at higher salinities, 15 and 30 ppt.

Cobia larvae metamorphose to gill respiration 11–15 days post hatching. At 15–25 days post hatching, cobia are weaned onto commercial formulated feeds. Rearing cobia larvae at salinities as low as 15 ppt is possible. Fully weaned fingerlings weighing up to one gram are transferred to juvenile culture tanks. Later cobia juveniles can be raised in ponds or shallow, near-shore submerged cages.

Juveniles thrive on a wide range of protein and lipid, but there are optimal levels where they get the most benefit. After an 8-week growth trial, juvenile cobia displayed a peak in their weight gain with a dietary protein concentration of 44.5%. Weight gain is also likely to increase as the lipid content in the diet increases. However, levels exceeding 15–18% produces little practical benefit because of higher fat accretion in the cobia. In addition, up to 40% of fish meal protein can be substituted with soybean meal protein before a reduction occurs in growth rates and protein utilization. Cobia has low feed conversion rates, yielding 1 kilogram of fish biomass for 1.8 kilograms of pellets which contain 50% fishmeal.

The cobia are then transferred to open ocean cages for final the grow-out when they reach 6–10 kilograms. The growth rate and survival rate of cobia during grow-out stages in open water cages throughout the Caribbean and Americas vary from as little as 10% up to 90%. Low survival rates are mainly due to disease, but also to shark attacks which tear holes in the nets of cages in the Bahamas and Puerto Rico and allow caged cobia to escape. However, better growth rates were experienced in offshore cage farms in Taiwan. In addition, cobia are considered to be gonochoristic, with differential growth rates occurring between sexes. Females grow faster and have been shown to be significantly longer and heavier within year classes.

Diseases

- Nephrocalcinosis (kidney stones) cause significant mortality during both the hatchery and grow-out stages. These stones vary in diameter from 2–6 mm in the kidney and can block the urethra. This condition is not fully understood, but is thought to be a symptom of prolonged exposure to free carbon dioxide in excess of 10 mg/L. The ratio of calcium to magnesium in the diet could also be out of balance.

- A *Sphaerospora*-like myxosporean infection caused 90% mortality during one month in a marine cage cultured in Taiwan.

Benefits and Constraints

Offshore aquaculture, regardless of the species, is beneficial because it can avoid conflict with recreational activities and local fisherman, as well as potentially improving the coastal aesthetics. Further, repositioning aquaculture facilities in less polluted open water environments can produce better products, and the high flushing rates experienced in the open ocean reduces the effect of effluents on benthic communities.

However, such operations require more developed infrastructure than near-shore aquaculture systems, which makes them expensive. Offshore sites have access difficulties and much higher labour costs.

Broodstock

An amberjack broodstock, *Seriola dumerili*

Broodstock, or broodfish, are a group of mature individuals used in aquaculture for breeding purposes. Broodstock can be a population of animals maintained in captivity as a source of replacement for, or enhancement of, seed and fry numbers. These are generally kept in ponds or tanks in which environmental conditions such as photoperiod, temperature and pH are controlled. Such populations often undergo conditioning to ensure maximum fry output. Broodstock can also be sourced from wild populations where they are harvested and held in maturation tanks before their seed is collected for grow-out to market size or the juveniles returned to the sea to supplement natural populations. This method, however, is subject to environmental conditions and can be unreliable seasonally, or annually. Broodstock management can improve seed quality and number through enhanced gonadal development and fecundity.

Management

Broodstock management involves manipulating environmental factors surrounding the broodstock to ensure maximum survival, enhance gonadal development and increase fecundity. Such conditioning is necessary to ensure the sustainability of aquaculture production. It is also utilised to increase the number and quality of eggs produced and control the timing of maturation and spawning. Management of the technologies for gamete production in captivity is one of the essential step for aquaculture that would ensure the growth to this sector. Unfortunately, most fish when reared in captivity condition, exhibit some degree of reproduction dysfunction. Many species of captive fish are able to reach reproduction maturity in aquaculture conditions and gonadal growth occurs normally. However, some of female species often fail final oocyte maturation stage. Hormonal manipulation and acceleration of final oocyte maturation due to the economics of broodstock management is important. For instance, in Salmoniformes, the need to collect the eggs by stripping is a serious limitation, while the time of ovulation must be predicted with accuracy, as over-ripening may take place in minutes or hours after ovulation Therefore, control of broodstock reproductive is essential for the sustainability of commercial aquaculture production.

Choosing species to use requires consideration of the biology of the species. This includes their size at maturity, method of reproduction, feeding behaviour and ability to tolerate adverse conditions Farms also consider whether they grow their own broodstock or obtain them from natural populations. Where natural populations are excluded, the farm can be considered a self-sustaining unit independent of external genetic influence.

Pond-reared broodstocks are selected, often as immature juveniles, and grown out in suitable conditions to sexual maturity. These animals require stable water characteristics and a well-balanced, species-dependent, protein rich diet. This enhances the germinal tissue for future seed stock as it is formed in juveniles.

The pond or tank in which broodfish are held must be a suitable size to hold and condition the broodstock. Dependent on the species involved you need to alter the number of individuals, and often separate the sexes. Sex separation enables the broodstock males and females to be subjected to different conditions where necessary. For example, male and female sturgeons respond to different hormone levels, this also allows more control over eggs and sperm.

The characteristics of the water in which the mature broodstocks are held must be manipulated. The aquaculturist must consider the appropriate oxygen concentration, temperature, and pH of the water all of which can be species specific.

The feeding regime of broodstocks is species specific and requires consideration of timing and composition of the food. Protein, lipid and fatty acid composition is particularly important. The quantity of food intake can be altered to influence spawning and maturity, for example low rations have been shown to reduce the number of fish reaching maturity while increasing the fecundity of those which do.

When fry are desired, spawning can be induced in broodstocks by manipulation of relevant environmental factors. In particular the photoperiod can be altered to imply that it is time to spawn. A shortended photoperiod is known to advance spawning times while a lengthened photoperiod can delay spawning. Artificial light can be used to change the apparent day length and indicate different seasonal features so as to delay spawning. Water temperature can be increased for the same purpose. Following spawning the female broodfish are often stressed and have lost weight. They require extra care and abundant feeding at this time to ensure survival to the next spawning season.

Advantages

Managers can select for reproductive characteristics which influence the egg producing capability of individuals and increase fecundity by providing them with optimal environment and diets. This is further possible in pond-reared populations where traits can be selected for over generations for example, for higher fecundity.

The breeding season and spawning times can be shifted thus expanding the seasonal range of production. This leads to more efficient aquaculture because fry are available to the market year round. Hormonal treatments can advance spawning by two to three weeks. Manipulating photoperiod can alter spawning time by over four months and is cheap and straightforward to achieve.

Broodstock managers can use or select for traits such as fast growth rates or disease resistance over generations to produce more desirable fish. This ability for genetic improvement of stocks is more efficient and produces higher value stock. Broodstocks also enable you to selectively plan and control all matings. Selective breeding is an important part of the domestication of aquaculture species.

Pond-reared broodstocks benefit from the removal of predation which can be a significant cause of mortality in natural populations. They further benefit from the removal of variable environmental impacts.

Holding broodstock in an accessible pond or tank offers readily available breeding adults whenever required.

Disadvantages

When broodstocks are used to supplement natural populations they face different selective pressures to normal. Thus they may not have adequate fitness to survive the natural environment, or can alter and decrease natural genetic diversity due to the bottleneck nature of breeding from a smaller population.

Broodstocks require supplementation from outside sources regularly to prevent negative effects of closed populations. Domestication of broodstocks in hatcheries can reduce reproductive capabilities and alter other genetic characteristics. For example, a trout stock maintained as a closed population for 20 generations showed reduced number and size of egg production.

Examples

Penaeidae

Shrimp, particularly of the family Penaeidae, are one of the largest internationally traded species. Native stocks are usually collected as sources of broodstock supply . There are also examples of pond-reared Penaeidae broodstocks. These shrimp are raised in suitable environmental conditions including a 12–14 hour/day photoperiod, a water temperature of 25–29 °C and full seawater salinity with high water exchange rates.

Sydney Rock Oyster

The Sydney rock oyster, *Saccostrea glomerata*, has been farmed in New South Wales, Australia for over 100 years. Due to declines in the supply in the past 30 years, New South Wales introduced a selection program in 1990 to breed faster growing stocks. The utilised broodstocks are held in artificial ponds of around 0.11 ha in size, and at low densities. Broodstocks provided higher numbers of larvae and could be spawned readily providing a more definite source of Sydney rock oysters.

Rainbow Trout

Global production of rainbow trout, *Oncorhynchus mykiss*, requires over 3 billion eggs per year. This number is met because of broodstocks which undergo selection and conditioning in hatcheries. Trout have been reared artificially for over 80 years. Rainbow trout broodstocks are commonly manipulated to delay maturation and spawning time in order to provide eggs regularly and optimise supply. Artificial selection has favoured larger fish due to evidence of correlations between fish size and fecundity.

Commercial Fish Feed

Fish meal factory, Bressay

Manufactured feeds are an important part of modern commercial aquaculture, providing the balanced nutrition needed by farmed fish. The feeds, in the form of granules or pellets, provide the nutrition in a stable and concentrated form, enabling the fish to feed efficiently and grow to their full potential.

Many of the fish farmed more intensively around the world today are carnivorous, for example Atlantic salmon, trout, sea bass, and turbot. In the development of modern aquaculture, starting in the 1970s, fishmeal and fish oil were key components of the feeds for these species. They are combined with other ingredients such as vegetable proteins, cereal grains, vitamins and minerals and formed into feed pellets. Wheat, for example, is widely used as it helps to bind the ingredients in the pellets.

Other forms of fish feed being used include feeds made entirely with vegetable materials for species such as carp, moist feeds preferred by some species (easier to make but more difficult to store), and trash fish — that is fish caught and fed directly to larger species being raised in aquaculture pens.

Hatchery Feeds

Specialised feeds are produced for fish hatcheries. In species such as salmon and trout, the newly hatched fry first feed from their yolk sacs and then can be fed with starter feeds. Marine species such as sea bass, sea bream, flounders and turbot consume the nutrition in their yolk sacs during the first few days post hatching and then are fed for several weeks on live prey, in the form of rotifers and brine shrimp (Artemia). Special feeds can be used to enrich the nutritional value of the prey. Rotifers are usually bred in the hatchery while brine shrimp are generally collected from the wild, e.g. salt lakes. Manufactured feed alternatives to brine shrimp are becoming available, offering more consistent nutrition and improved sustainability as demand increases with the growth of aquaculture.

Development of Manufactured Feeds

Until the end of World War II most fish hatcheries relied on raw meat (horse meat in particular) as a dietary staple for trout. In the early 1950s, John E. (Red) Hanson, while working for the New Mexico Game and Fish Department, began experimenting with dietary routine and dry pellet formulations. The first fish feed pellets were introduced to hatchery trout at the Red River Hatchery near Taos. The pellets resulted in improved conversion rates of food intake to fish production, and lead in turn to the wider adoption of fish pellets in hatcheries.

The development of dry pelleted fish feeds to date has two themes. One theme is on improving digestibility and refining the balance of nutrients to match the needs of the different species of fish more precisely at different periods of development. The other theme is to improve the sustainability of the ingredients used. This is being achieved mainly by identifying additional sustainable sources of ingredients, in particular to reduce the need for fishmeal and fish oil. Improving the efficiency of feeding also contributes to sustainability.

Sustainability

Fish feed production in Stokmarknes Norway

Traditionally two of the most important ingredients have been fishmeal and fish oil. These come mainly from the processing of fish from the wild catch, usually pelagic species that are generally not suited to processing for human consumption. Fish sold for human consumption attract a higher price than those used to make fishmeal. The fishmeal fisheries are often referred to as reduction fisheries. The world's largest reduction fishery is in the Pacific, off the coast of Peru and Chile and is regulated by the governments of those countries. The North Atlantic is another important source of fish for fishmeal and fish oil. Many major suppliers belong to the International Fishmeal and Fish Oil Organisation.

Dry fish feed pellets

Fishmeal is a brown, flour-like material made by specialist producers that cook, press, dry and grind the fish. The fish oil is effectively a by-product of this process that proves to be a rich source of energy and fatty acids for fish, including the important long-chain omega-3 fatty acids EPA and DHA now linked to the health benefits associated with eating oily fish such as salmon and mackerel. Fish in general also are good sources of many vitamins and minerals and are often recommended as part of a healthy diet by governmental food agencies.

Because the catches of wild fish must be managed at sustainable levels to ensure the stocks continue to be viable, the available supply of fishmeal and fish oil from these resources will not increase.

The global demand for fish from consumers around the world is increasing. Reasons include the growing population, rising average incomes and greater awareness of fish as part of a healthy diet. The yield from the wild catch cannot be increased sustainably, therefore, in the opinion of observers such as the Food and Agriculture Organization (FAO) of the United Nations, aquaculture must fill the gap. Currently the supply of fish from aquaculture approximately matches that from the wild catch, according to FAO figures.

The current drive in research and development is enabling this to happen by supplementing fishmeal and fish oil with vegetable proteins and oils, while ensuring the fish continue to provide the important health benefits for consumers. Other potential raw material resources are also being explored. For example, the U.S. biotechnology company BioTork is piloting the use of raw materials such as unmarketable papaya and by-products from biodiesel production to produce fish feed components, as well as feeding agricultural waste to algae and fungi that manufacture some of the proteins and omega-3 oils needed for fish food. The US biotechnology company Calysta and the UK/Danish biotech company Unibio opened small plants in the UK and Denmark to produce fish feed from natural gas in 2016.

Modern Fish Feed

Modern fish feeds are made by grinding and mixing together ingredients such as fishmeal, vegetable proteins and binding agents such as wheat. Water is added and the resulting paste is extruded through holes in a metal plate. The diameter of the holes sets the diameter of the pellets, which can range from less than a millimetre to over a centimetre. As the feed is extruded it is cut to form pellets of the required length. The pellets are dried and oils are added. Adjusting parameters such as temperature and pressure enables the manufacturers to make pellets that suit different fish farming methods, for example feeds that float or sink slowly and feeds suited to recirculation systems. The dry feed pellets are stable for relatively long periods, for convenient storage and distribution. Feeds are delivered in bulk, in large bags—usually one tonne, or in 25 kilogram bags. Smaller quantities of specialist feeds are supplied for use in fish hatcheries. The three major manufacturers of fish feeds for aquaculture are Biomar, EWOS and Skretting.

Extruded fish feed

Fish Hatchery

Tanks in a shrimp hatchery.

A fish hatchery is a "place for artificial breeding, hatching and rearing through the early life stages of animals, finfish and shellfish in particular". Hatcheries produce larval and juvenile fish (and shellfish and crustaceans) primarily to support the aquaculture industry where they are transferred to on-growing systems i.e. fish farms to reach harvest size. Some species that are commonly raised in hatcheries include Pacific oysters, shrimp, Indian prawns, salmon, tilapia and scallops. The value of global aquaculture production is estimated to be US$98.4 billion in 2008 with China significantly dominating the market, however the value of aquaculture hatchery and nursery production has yet to be estimated. Additional hatchery production for small-scale domestic uses, which is particularly prevalent in South-East Asia or for conservation programmes, has also yet to be quantified.

There is much interest in supplementing exploited stocks of fish by releasing juveniles that may be wild caught and reared in nurseries before transplanting, or produced solely within a hatchery. Culture of finfish larvae has been utilised extensively in the United States in stock enhancement efforts to replenish natural populations. The U.S. Fish and Wildlife Service have established a National Fish Hatchery System to support the conservation of native fish species.

Purpose

Assynt Salmon hatchery, near Inchnadamph in the Scottish Highlands.

Hatcheries produce larval and juvenile fish and shellfish for transferral to aquaculture facilities where they are 'on-grown' to reach harvest size. Hatchery production confers three main benefits to the industry:

1. Out of season production

Consistent supply of fish from aquaculture facilities is an important market requirement. Broodstock conditioning can extend the natural spawning season and thus the supply of juveniles to farms. Supply can be further guaranteed by sourcing from hatcheries in the opposite hemisphere i.e. with opposite seasons.

2. Genetic improvement

Genetic modification is conducted in some hatcheries to improve the quality and yield of farmed species. Artificial fertilisation facilitates selective breeding programs which aim to improve production characteristics such as growth rate, disease resistance, survival, colour, increased fecundity and/or lower age of maturation. Genetic improvement can be mediated by selective breeding, via hybridization, or other genetic manipulation techniques.

3. Reduce dependence on wild-caught juveniles

In 2008 aquaculture accounted for 46% of total food fish supply, around 115 million tonnes. Although wild caught juveniles are still utilised in the industry, concerns over sustainability of extracting juveniles, and the variable timing and magnitude of natural spawning events, make hatchery production an attractive alternative to support the growing demands of aquaculture.

Production Steps

Manually stripping eggs

Juvenile salmon towards the end of their stay in a hatchery

Broodstock

Broodstock conditioning is the process of bringing adults into spawning condition by promoting the development of gonads. Broodstock conditioning can also extend spawning beyond natural spawning periods, or for production of species reared outside their natural geographic range with different environmental conditions. Some hatcheries collect wild adults and then bring them in for conditioning whilst others maintain a permanent breeding stock. Conditioning is achieved by holding broodstock in flow-through tanks at optimal conditions for light, temperature, salinity, flow rate and food availability (optimal levels are species specific). Another important aspect of broodstock conditioning is ensuring the production of high quality eggs to improve growth and survival of larvae by optimising the health and welfare of broodstock individuals. Egg quality is often determined by the nutritional condition of the mother. High levels of lipid reserves in particular are required to improve larval survival rates.

Spawning

Natural spawning can occur in hatcheries during the regular spawning season however where more control over spawning time is required spawning of mature animals can be induced by a variety of methods. Some of the more common methods are:

Manual stripping : For shellfish, gonads are generally removed and gametes are extracted or washed free. Fish can be manually stripped of eggs and sperm by stroking the anaesthetised fish under the pectoral fins towards the anus causing gametes to freely flow out.

Environmental manipulation: Thermal shock, where cool water is alternated with warmer water in flow-through tanks can induce spawning. Alternatively, if environmental cues that stimulate natural spawning are known, these can be mimicked in the tank e.g. changing salinity to simulate migratory behaviour. Many individuals can be induced to spawn this way, however this increases the likelihood of uncontrolled fertilisation occurring.

Chemical injection: A number of chemicals can be used to induce spawning with various hormones being the most commonly used.

Fertilisation

Prior to fertilisation, eggs can be gently washed to remove wastes and bacteria that may contaminate cultures. Promoting cross-fertilisation between a large number of individuals is necessary to retain genetic diversity in hatchery produced stock. Batches of eggs are kept separate, fertilised with sperm obtained from several males and allowed to stand for an hour or two before samples are analyzed under a microscope to ensure high rates of fertilisation and to estimate numbers to be transferred to larval rearing tanks.

Larvae

Rearing larvae through the early life stages is conducted in nurseries which are generally closely associated with hatcheries for fish culture whilst it is common for shellfish nurseries to exist separately. Nursery culture of larvae to rear juveniles of a size suitable for transferral to on-growing facilities can be performed in a variety of different systems which may be entirely land-based, or larvae may be later transferred to sea-based rearing systems which reduce the need to supply feed. Juvenile survival is dependent on very high quality water conditions. Feeding is an important component of the rearing process. Although many species are able to grow on maternal reserves alone (lecithotrophy), most commercially produced species require feeding to optimise survival, growth, yield and juvenile quality. Nutritional requirements are species specific and also vary with larval stage. Carnivorous fish are commonly fed with live prey; rotifers are usually offered to early larvae due to their small size, progressing to larger *Artemia* nauplii or zooplankton. The production of live feed on-site or buying-in is one of the biggest costs for hatchery facilities as it is a labour-intensive process. The development of artificial feeds is targeted to reduce the costs involved in live feed production and increase the consistency of nutrition, however decreased growth and survival has been found with these alternatives.

Settlement of Shellfish

The hatchery production of shellfish also involves a crucial settling phase where free-swimming larvae settle out of the water onto a substrate and undergo metamorphosis if suitable conditions are found. Once metamorphosis has taken place the juveniles are generally known as spat, it is this phase which is then transported to on-growing facilities. Settlement behaviour is governed by a range of cues including substrate type, water flow, temperature, and the presence of chemical cues indicating the presence of adults, or a food source etc. Hatchery facilities therefore need to understand these cues to induce settlement and also be able to substitute artificial substrates to allow for easy handling and transportation with minimal mortality.

Hatchery Design

Multi-Species Fish and Invertebrate Breeding and Hatchery, (Oceanographic Marine Laboratory in Lucap, Alaminos, Pangasinan, Philippines, RMaTDeC,2011).

Hatchery designs are highly flexible and are tailored to the requirements of site, species produced, geographic location, funding and personal preferences. Many hatchery facilities are small and coupled to larger on-growing operations, whilst others may produce juveniles solely for sale. Very small-scale hatcheries are often utilized in subsistence farming to supply families or communities particularly in south-east Asia. A small-scale hatchery unit consists of larval rearing tanks, filters, live food production tanks and a flow through water supply. A generalized commercial scale hatchery would contain a broodstock holding and spawning area, feed culture facility, larval culture area, juvenile culture area, pump facilities, laboratory, quarantine area, and offices and bathrooms.

Expense

Labour is generally the largest cost in hatchery production making up more that 50% of total costs. Hatcheries are a business and thus economic viability and scale of production are vital considerations. The cost of production for stock-enhancement programmes is further complicated by the difficulty of assessing the benefits to wild populations from restocking activities.

Issues

Genetic

Hatchery facilities present three main problems in the field of genetics. The first is that maintenance of a small number of broodstock can cause inbreeding and potentially lead to inbreeding depression thus affecting the success of the facility. Secondly, hatchery reared juveniles, even from a fairly large broodstock, can have greatly reduced genetic diversity compared to wild populations (the situation is comparable to the founder effect). Such fish that escape from farms or are released for restocking purposes may adversely affect wild population genetics and viability. This is of particular concern where escaped fish have been actively bred or are otherwise genetically modified. The third key issue is that genetic modification of food items is highly undesirable for many people.

Fish Farms

Other arguments that surround fish farms such as the supplementation of feed from wild caught species, the prevalence of disease, fish welfare issues and potential effects on the environment are also issues for hatchery facilities.

References

- Avnimelech, Y; Kochva, M; et al. (1994). "Development of controlled intensive aquaculture systems with a limited water exchange and adjusted carbon to nitrogen ratio.". Israeli Journal of Aquaculture Bamidgeh. 46 (3): 119–131

- Aquaculture, Office of. "Basic Questions about Aquaculture :: Office of Aquaculture". www.nmfs.noaa.gov. Retrieved 2016-06-09

- Torrissen, Ole; et al. (2011). "Atlantic Salmon (Salmo Salar): The 'Super-Chicken' Of The Sea?.". Reviews In Fisheries Science. 19 (3): 257–278. doi:10.1080/10641262.2011.597890

- "Current issues in fish welfare". Journal of Fish Biology. 68 (2): 332–372. 2006. doi:10.1111/j.0022-1112.2006.001046.x

- Barrionuevo, Alexei (July 26, 2009). "Chile's Antibiotics Use on Salmon Farms Dwarfs That of a Top Rival's". The New York Times. Retrieved 2009-08-28

- Weaver, D E (2006). "Design and operations of fine media fluidized bed biofilters for meeting oligotrophic water requirements". Aquacultural Engineering. 34 (3): 303–310. doi:10.1016/j.aquaeng.2005.07.004

- Faulk, C.K. & Holt, G.J. (2003) Lipid nutrition and feeding of cobia Rachycentron canadum larvae Journal of the World Aquaculture Society, 34: 368–378

- Shore, Randy (17 November 2012) Salmon farming comes ashore in land-based aquaculture The Vancouver Sun, Retrieved 21 February 2013

- Krkošek, Martin; et al. (2007). "Report: "Declining Wild Salmon Populations in Relation to Parasites from Farm Salmon". Science. 318 (5857): 1772–1775. PMID 18079401. doi:10.1126/science.1148744

- Denson, M.R., Stuart, K.R., Smith, T.I.J., Weirich, C.R. & Segars, A. (2003) "Effects of salinity on growth, survival, and selected hematological parameters of juvenile cobia Rachycentron canadum" Journal of the World Aquaculture Society, 34: 496–504

- Naylor, RL; Goldburg, RJ; Mooney, H; et al. (1998). "Nature's Subsidies to Shrimp and Salmon Farming". Science. 282 (5390): 883–884. doi:10.1126/science.282.5390.883

- Yousefian, M., Mousavi, S. E.(2011). A review of the control of reproduction and hormonal manipulations in finfish species. African Journal of Agricultural Research, 6(7), 1643-1650

Permissions

Index

www.ingramcontent.com/pod-product-compliance
Lightning Source LLC
Chambersburg PA
CBHW082100190326
41458CB00010B/3530